2013—2025年国家辞书编纂出版规划项目
英汉信息技术系列辞书

总主编　白英彩

A CONCISE ENGLISH-CHINESE DICTIONARY OF INTELLIGENT BUILDING TECHNOLOGY

英汉建筑智能化技术简明词典

主　编　沈忠明　曾松鸣
主　审　赵哲身
副主编　沈　晔　王明敏　董莹荷

内容提要

本词典为“英汉信息技术系列辞书”之一。本词典收录了建筑智能化技术及其产业领域相关的理论研究、开发应用、工程管理等方面的专业词汇 8 000 余条，所有词汇均按照英文字母顺序排列，对所有收录的词条进行了梳理、规范和审定。

本词典可供建筑智能化等相关领域从事研究、开发和应用的人员、信息技术书刊编辑和文献译摘人员使用，也适合相关专业的大专院校师生参考。

图书在版编目(CIP)数据

英汉建筑智能化技术简明词典/沈忠明，曾松鸣主编. —上海：上海交通大学出版社，2021
ISBN 978-7-313-23226-7

Ⅰ. ①英… Ⅱ. ①沈… ②曾… Ⅲ. ①智能化建筑—词典—英、汉 Ⅳ. ①TU855-61

中国版本图书馆 CIP 数据核字(2020)第 077569 号

英汉建筑智能化技术简明词典

YINGHAN JIANZHU ZHINENGHUA JISHU JIANMING CIDIAN

主　　编：沈忠明　曾松鸣
出版发行：上海交通大学出版社　　地　　址：上海市番禺路 951 号
邮政编码：200030　　电　　话：021-64071208
印　　制：苏州市越洋印刷有限公司　　经　　销：全国新华书店
开　　本：880 mm×1230 mm　1/32　　印　　张：8.25
字　　数：240 千字
版　　次：2021 年 4 月第 1 版　　印　　次：2021 年 4 月第 1 次印刷
书　　号：ISBN 978-7-313-23226-7
定　　价：98.00 元

英汉信息技术系列辞书顾问委员会

英汉信息技术系列辞书编纂委员会

英汉建筑智能化技术简明词典

编　委　会

序

“信息技术”(IT)这个词如今已广为人们知晓,它通常涵盖计算机技术、通信(含移动通信)技术、广播电视技术、以集成电路(IC)为核心的微电子技术和自动化领域中的人工智能(AI)、神经网络、模糊控制和智能机器人,以及信息论和信息安全等技术。

近20多年来,信息技术及其产业的发展十分迅猛。20世纪90年代初,由信息高速公路掀起的IT浪潮以来,信息技术及其产业的发展一浪高过一浪,因特网(互联网)得到了广泛的应用。如今,移动互联网的发展势头已经超过前者。这期间还涌现出了电子商务、商务智能(BI)、对等网络(P2P)、无线传感网(WSN)、社交网络、网格计算、云计算、物联网和语义网等新技术。与此同时,开源软件、开放数据、普适计算、数字地球和智慧地球等新概念又一个接踵一个而至,令人应接不暇。正是由于信息技术如此高速的发展,我们的社会开始迈入“新信息时代”,迎接“大数据”的曙光和严峻挑战。

如今信息技术,特别是“互联网+”已经渗透到国民经济的各个领域,也贯穿到我们日常生活之中,可以说信息技术无处不在。不管是发达国家还是发展中国家,人们之间都要互相交流,互相促进,缩小数字鸿沟。

上述情形映射到信息技术领域是:每年都涌现出数千个新名词、术语,且多源于英语。编纂委认为对这些新的英文名词、术语及时地给出恰当的译名并加以确切、精准的理解和诠释是很有意义的。这项工作关系到IT界的国际交流和大陆与港、澳、台之间的沟通。这种交流不限于学术界,更广泛地涉及IT产业界及其相关的商贸活动。更重要的是,这项工作还是IT技术及其产业标准化的基础。

编纂委正是基于这种认识，特组织众多专家、学者编写《英汉信息技术大辞典》《英汉计算机网络辞典》《英汉计算机通信辞典》《英汉信息安全技术辞典》《英汉三网融合技术辞典》《英汉人工智能辞典》《英汉建筑智能化技术辞典》《英汉智能机器人技术辞典》《英汉智能交通技术辞典》《英汉云计算·物联网·大数据辞典》《英汉多媒体技术辞典》和《英汉微电子技术辞典》，以及与这些《辞典》（每个词汇均带有释文）相对应的《简明词典》（每个词汇仅有中译名而不带有释文）共24册，陆续付梓。我们希望这些书的出版对促进IT的发展有所裨益。

这里应当说明的是编写这套书籍的队伍从2004年着手，预计将历时17年完成，与时俱进的辛勤耕耘，终得硕果。他们早在20世纪80年代中期就关注这方面的工作并先后出版了《英汉计算机技术大辞典》（获得中国第十一届图书奖）及其类似的书籍，参编人数一直持续逾百人。虽然参编人数众多，又有些经验积累，但面对IT技术及其产业化如此高速发展，相应出现的新名词、术语之多，尤令人感到来不及收集、斟酌、理解和编纂之虞。如今推出的这套辞书不免有疏漏和欠妥之处，请读者不吝指正。

这里，编纂委尤其要对众多老专家执着与辛勤耕耘表示由衷的敬意，没有他们对事业的热爱，没有他们默默奉献的精神，没有他们追求卓越的努力，是不可能成就这一丰硕成果的。

在“英汉信息技术系列辞书”编辑、印刷、发行各个环节都得到上海交通大学出版社大力支持。尤其值得我们欣慰的是由上海交通大学和编纂委共同聘请的15位院士和多位专家所组成的顾问委员会对这项工作自始至终给予高度关注、亲切鼓励和具体指导，在此也向各位资深专家表示诚挚谢意！

编纂委真诚希望对这项工作有兴趣的专业人士给予支持、帮助并欢迎加盟，共同推动该工程早日竣工，更臻完善。

英汉信息技术系列辞书编纂委员会

名誉主任：吴启迪

2015年5月18日

前　言

随着信息科学和技术迅速发展和我国现代化建设的推进，现代信息技术在建筑领域的应用日益广泛和深入，智能家居、智能建筑、智慧社区、智慧城市得到极大发展和普及，建筑智能化以人、建筑、环境互为协调出发，集架构、系统、应用、管理和优化组合为一体，充分应用计算机、通信、物联网、自动控制、人工智能等现代信息技术于建筑的建设和运营之中，成为现代建筑不可或缺的重要组成部分，为人们提供安全、高效、便利和可持续发展的功能环境，并深刻改变着人们的思维、生活、生产方式，推动着社会经济各个领域的飞速发展。

自20世纪末开始，建筑智能化技术和产品从引进、应用到自主研发和生产，已经历了30余年时间，随着“互联网+”的驱动和导向，建筑智能化领域的学习、研究、应用和交流必将持续加强和深化。为适应建筑智能化迅速发展及新兴的智能建筑行业业务活动的实际需求，我们编纂出版《英汉建筑智能化技术简明词典》，期望为建筑智能化及相关的研究人员、工程技术人员、操作使用人员和相关的管理人员提供便利。

本词典收录了建筑智能化相关的规范化名词术语8 000余条，内容涵盖了信息设施系统、建筑设备管理系统、公共安全系统、智能化集成系统、电子信息机房、信息化应用系统等领域的名词术语。

本词典自2014年起组织编写，得到上海市智能建筑建设协会和编写组人员所在单位的大力支持。编写组人员分工汇集初稿形成后，在沈晔、王明敏、董莹荷等三位同志的协助下，由曾松鸣同志帮助进一步仔细地梳理、查核和汇总，并交付赵哲身教授等主审。他们一丝不苟，精益求精，逐条审核了全书。系列辞书总主编白英彩教授对本词典的编纂给予了细致而具体的指导。这里谨向各位同仁、专家致以诚挚的谢意！由于建筑智能化技术和产业发展迅速，新的名词、术语不断涌现，加之编者水平有限，存在疏漏和不当之处，恳请各位读者不吝指教。

感谢深圳市普联技术公司董事长赵建军先生和上海金桥信息股份有限公司、广州宇洪科技股份有限公司对本书出版的支持和鼎力资助！

沈忠明　谨识

凡　　例

1. 本词典按英文字母顺序排列，不考虑字母大小写。数字和希腊字母另排。专用符号(圆点、连字符等)不参与排序。
2. 英文词汇及其对应的中文译名用粗体。一个英文词汇有几个译名时，可根据彼此意义的远近用分号或者逗号隔开。
3. 圆括号“(　)”内的内容表示解释或者可以略去。如“connecting hardware 连接(硬)件[连接器(硬)件]”；也可表示某个词汇的缩略语，如 above ground structure (AGS)。
4. 方括号“[　]”内的内容表示可以替换紧挨方括号的字词。如“cabling 布线[缆]”。
5. 单页码上的书眉为本页最后一个英文词汇的第一个单词；双页码上的书眉为本页英文词汇的第一个单词。
6. 英文名词术语的译名以全国科学技术委员会名词审定委员会发布的为主要依据，对于已经习惯的名词也作了适当反映，如“disk”采用“光碟”为第一译名，“光盘”为第二译名。
7. 本词典中出现的计量单位大部分采用我国法定计量单位，但考虑到读者查阅英文技术资料的方便，保留了少量英制单位。

广州宇洪科技股份有限公司简介

广州宇洪科技股份有限公司(简称“宇洪科技”)成立于2007年11月,总部位于广州市天河区软件路11号,智慧城核心区国家软件(广州)产业基地内。公司致力于让信号传输变得更融合、更畅通、更可靠,是以信号传输为核心技术的物联网解决方案与服务提供商,专注于物联网的“传感+传输+智慧”系统研究。研发实力雄厚,率先开发了服务于智慧城市的二维码可视化标签管理系统,在资产管理、路由管理、链路告警、维护运营等方面均处于行业领先地位。公司于2017年挂牌新三板,股票代码:872199。

宇洪科技面向智能建筑、轨道交通、智慧城市等行业应用提供信号传输端到端的整体解决方案,包括产品销售、业务咨询与信号传输项目集成、软硬件系统集成等工程服务业务。先后被评为广州市著名商标、国家高新技术企业、广州科技小巨人企业等。公司产品多次被行业推荐为“民族品牌”“行业十大品牌”等。先后通过泰尔、UL、CCC等多家专业机构的认证。

宇洪科技已经在全国设有四大分支机构、九大营销中心和三十多个办事处,业务辐射全国,同时积极拓展海外业务,已在马来西亚、新加坡、越南等地设有海外分公司。

上海金桥信息股份有限公司简介

上海金桥信息股份有限公司(SH.603918)创立于1994年,2015年5月于上海证券交易所挂牌上市。经近三十年稳健发展,公司业务体系完整,并覆盖全国绝大多数省(市)及地区,设有二十多个分支机构。公司主营业务是为客户提供智慧空间信息化解决方案及服务。公司秉承"真诚是金,共享为桥"的核心价值观,以"服务人与人、人与信息、人与环境之间的沟通"为主线,充分发挥信息化领域综合优势,面向政务、司法、教育、医疗健康、金融等行业及各类企业,融合客户需求,叠加行业应用,打造智慧空间信息化系统系列解决方案,以卓越技术实力与全心全意支持与服务理念,在业内赢得良好口碑与声誉。

公司大力推行人才战略与创新发展战略,不断加大研发投入,打造核心竞争力,努力保持行业领先。公司拥有一批注册建造师,信息系统集成及服务中级、高级工程师,中级、高级项目经理等行业专业人才资源。公司已拥有建筑智能化系统设计专项甲级、电子与智能化工程专业承包一级、信息系统建设和服务能力等级CS4等资质。先后已实施近万项信息工程项目,获得行业权威机构认可,荣获中国建筑工程"鲁班奖"(参与浦东新区办公中心工程)、上海建设工程"白玉兰"奖(参与上海市委组织部、宣传部、人事局办公大楼工程)、上海市智能建筑"申慧奖"(上海世博中心多媒体会议系统工程等)国家和行业荣誉。行业历次评优均被评为"上海市智能建筑设计施工优秀企业",2020年通过上海市"市级企业技术中心"认证。

目　录

A

a data communication protocol for building automation and control networks (BACnet) 楼宇自动控制网络数据通信协议

A/D (analog/digital) 模拟/数字

A/D (analog-to-digital conversion) 模数转换

A/D and D/A combined converter A/D与D/A组合转换器

A/D converter (ADC) A/D 转换器,模数转换器

A/V (audio/video) 音视频

AAA (authentication, authorization, accounting) 认证,授权,计费

AAA sever AAA 服务器

AAC (adaptive audio coding) 自适应音频编码,高级音频编码

AACR-F (attenuation to alien/exogenous crosstalk ratio at the far-end) 远端衰减与外部串扰比

AACR-N (attenuation to alien/exogenous crosstalk ratio at the near-end) 近端衰减与外部串扰比

AAF (advanced authoring format) 先进制作格式

AAL (ATM adaptation layer) 异步传输模式[ATM]适配层

AAW (aluminum alloy wire) 铝合金线

abamurus 扶壁(外墙突出之墙垛)

aberration 像差

ABF (air blast freezing) 强制通风冻结

ABF (auto back focus) 自动后焦调节

abnormal environmental condition 异常环境条件

above ground structure (AGS) 地面建筑物

ABS (acrylonitrile butadiene styrene) 丙烯腈-丁二烯-苯乙烯

absent extension advice 无人分机通报器

absent extension diversion 分机用

A

户缺席转接

absolute humidity 绝对湿度

absolute permeability 绝对磁导率

absolute temperature scale (ATS) 绝对温标

absorption peak 吸收峰值

absorptive coefficient of solar radiation 太阳能辐射吸收系数

abstract syntax notation one (ASN.1) 抽象描述语言(一种 ISO/ITU-T 标准)

abutment 对接

AC (access code) 访问码

AC (access control) 访问控制

AC (adaptive control) 自适应控制

AC (alternating current) 变流电

AC input power 交流输入电源

AC powered embedded thermal control equipment 交流电源嵌入式温控设备

AC-3 (audio coding generation 3) 环绕声数字音频编码

ACC (air cooled condenser) 风冷冷凝器

accelerated graphical port (AGP) 图形加速端口

acceptance group 验收小组

acceptance inspection 竣工验收

acceptance test 验收测试

acceptance testing 验收测试

access attendance 门禁考勤

access card 门禁卡

access code (AC) 访问码

access control (AC) 访问控制

access control list (ACL) 访问控制列表

access control system (ACS) 出入口控制系统

access controller 门禁控制器

access gateway of IoT 物联网接入网关

access group 门禁组

access log 访问日志

access mainframe 门禁控制主机

access network (AN) 接入网

access path 接入通道

access point (AP) (无线网的)接入点,访问点

access privilege 访问特权,接入特许

access provider 接入提供商

access system 门禁系统

accessory building 辅助建筑物

accessory module 配套组件

accident analysis 事故分析

accident analysis report 事故调查报告

accident book 意外事故记录册

accident control 事故控制

accident frequency 事故频率

accident hazard 事故危险
accident investigation 事故调查
accident pattern 事故类型
accident prevention 事故预防
accident prevention program (APP) 事故预防计划
accident probability 事故可能性
accident rate 事故率
accident recorder 事故记录器
accident report 事故报告
accident spot 事故现场
accident statistic 事故统计
accident voucher of tooling 工艺装备事故报告单
accidental rate analysis 事故率分析
accidental report 事故报告(书)
accidental risk 事故危险性
accidental severity 事故严重程度
accidental site 失事地点
accommodation 住宿
accommodation stairway 简易楼梯
account 账户
accreditation 合格鉴定
ACCU (air cooled condensing unit) 风冷冷凝机组
ACD (automatic call distribution) 自动话务分配[分配话务]
ACD (automatic call distributor) 自动呼叫分配器
ACE (air conditioning equipment) 空气调节设备
ACE (auxiliary control element) 辅助控制单元
ACF (activity controlled frame rate) 活动状态帧率控制
ACF (adaptive comb filter) 自适应梳状滤波器
acid wash 酸洗
acknowledged information 确认信息
acknowledged information transfer service (AITS) 确认信息传递服务
ACL (access control list) 访问控制列表
acoustic amplification system 扩声系统
acoustic echo canceler (AEC) 回声消除器
acoustic power level (APL) 声功率级
ACP (air conditioning process) 空气调节过程
acquisition 采集
ACR (attenuation to crosstalk ratio) 衰减串扰比
ACR-F (attenuation to crosstalk ratio at the far-end) 远端衰减与串扰比

A

ACR-N (attenuation to crosstalk ratio at the near-end) 近端衰减与串扰比

acrylonitrile styrene acrylate copolymer (ASA) 丙烯腈-苯乙烯-丙烯酸共聚物

acrylonitrile butadiene styrene (ABS) 丙烯腈-丁二烯-苯乙烯

ACS (access control system) 出入口控制系统

ACS (advanced connectivity system) 先进布线系统(美国 IBM 公司综合布线系统的品牌名称)

ACS (air conditioning system) 空气调节系统

action signal 动作信号

action with alarm 报警联动

active 3D 主动 3D

active alarm 自动报警器

active device 有源器件

active directory service (ADS) 活动目录服务

active electronic circuitry 有源电子电路

active equipment 有源设备

active Ethernet 有源以太网,以太网供电

active format descriptor (AFD) 有效格式描述符

active infrared detector 主动红外探测器

active infrared intrusion detector 主动红外入侵探测器

active matrix 有源矩阵

active power 有功功率

active speaker 有源音箱

activity controlled frame rate (ACF) 活动状态帧率控制

activity detection 活动侦测

ACTS (advanced communication technology satellite) 高级通信技术卫星

actuator 执行器

adaptation layer ATM (AAL) ATM 适配层

adapter 适配器

adaptive audio coding (AAC) 自适应音频编码

adaptive comb filter (ACF) 自适应梳状滤波器

adaptive control (AC) 自适应控制

adaptive differential pulse code modulation (ADPCM) 自适应差分脉冲编码调制

ADC (A/D converter) A/D 转换器,模数转换器

add & drop multiplexer (ADM) 分插复用器

add/drop 分插,分出-插入,分路-插入

add/drop applications 分插方式应用

add/drop multiplexer (ADM) 分插复用器

add-in program 附加程序

additional building 附加建筑物

additional charge 附加费

additional factor for exterior door 外门附加率

additional factor for wind force 风力附加率

additional heat loss 附加耗热量

additional load 附加载荷

additional service 附加业务

add-on security 附加安全措施

addressable detector 可编址探测器

addressing 寻址

adhesive power 附着力

adhesive strip 黏合带

adjustable block 调节块

adjusting valve 调节阀

adjustment certificate 调校证书

ADM (add & drop multiplexer) 分插复用器

ADM (add/drop multiplexer) 分插复用器

administration 管理

administration module 管理模块

administration unit (AU) 管理单元

administration unit pointer (AU-PTR) 管理单元指针

administrative area 行政管理区

administrative unit alarm indication signal (AU-AIS) 管理单元告警指示信号

administrative unit group (AUG) 管理单元组

administrator 主管

ADP (apparatus dew point) 仪器露点

ADPCM (adaptive differential pulse code modulation) 自适应差分脉冲编码调制

ADS (active directory service) 活动目录服务

ADSL (asymmetrical digital subscriber line) 非对称数字用户线路

ADSL (asymmetrical digital subscriber loop) 非对称数字用户环路

ADSL transmission unit-central (ATU-C) ADSL 传输中央单元(Modem)

ADSL transmission unit-remote (ATU-R) ADSL 远传单元(Modem)

ADSL-high-speed internet access ADSL-高速因特网接入

ADSS (all dielectric self-support) 全介质自承式光缆

A

advanced audio coding (AAC) 高级音频编码

advanced authoring format (AAF) 先进制作格式

advanced communication technology satellite (ACTS) 高级通信技术卫星

advanced connectivity system (ACS) 先进布线系统(美国IBM公司综合布线系统的品牌名称)

advanced video coding (AVC) 高级视频编码

adverse slope 反坡

advice of charge (AoC) 收费通知

ADX (average directional index) 平均趋向指数

AEC (acoustic echo canceler) 回声消除器

AEER (annual energy efficiency ratio) 全年能效比

aerial cable 架空电缆

aerial fiber optic cable 架空光缆

aerial insulated cable 架空绝缘电缆

aerial ladder fire truck 云梯消防车

AES (automatic electronic shutter) 自动电子快门

AESS (automatic explosion suppression system) 自动抑爆系统

AEV (automatic expansion valve) 自动膨胀阀

AFD (active format descriptor) 有效格式描述符

AFD (atmospheric freeze drying) 常压冷冻干燥

AFDS (automatic fire detection system) 自动火灾探测系统

AFEXT (alien/exogenous far-end crosstalk loss) 外部远端串扰损耗

AFR (air filter regulator) 空气过滤调节器

AFR (amplitude-frequency response) 幅频响应

after flow 塑性变形

after service 售后服务

afterburner 补燃器,复燃室

after-condenser 后冷凝器,二次冷凝器

aftercooler 后冷却器,二次冷却器

after-flaming 补充燃烧

afterheat 余热

after-sales maintenance and organization plan 售后维护组织计划

after-sales service 售后服务

AFW (air flow window) 通风窗

AGC (automatic gain control) 自动增益控制

age hardening 老化,时效硬化
agent 自主体
aging 老化
aging of material 材料老化
agitator 搅拌器
AGP (accelerated graphical port) 图形加速端口
AGS (above ground structure) 地面建筑物
AHU (air handling unit) 空气处理单元[机组],空调箱
AI (analog input) 模拟输入,模拟量输入
AI (artificial intelligence) 人工智能
AIM (automated infrastructure management) 自动化基础设施管理
air accumulation 集气,聚气
air admitting surface 进风口面积
air anion generator 空气负离子发生器
air balance 风量平衡
air barrel 空气室
air blanket 空气夹层(气垫橡皮布的一种俗称)
air blast 鼓风,喷气器,气喷净法,气流
air blast connection pipe 风管,高压空气导管
air blast cooling 吹风式冷却
air blast freezing (ABF) 强制通风冻结
air blast freezing plant 强制通风冻结装置
air bleed hole 排气孔
air bleeder 放气管
air blender 空气混合器
air blower 鼓风机
air bottle 压缩空气瓶
air box 空气箱
air brake 气压制动器
air capacitor 空气电容器
air change 换气
air change rate (通风)换气率
air changes 换气次数
air circuit breaker 空气断路器
air circulation 空气循环
air classifier 气流分级机
air cleaner 空气滤清器
air cleaning 空气净化
air cleaning device 空气清洁装置
air compressor 空气压缩机
air condenser 空气冷凝器
air condition 空气调节
air condition system project 空调系统工程
air condition terminal 空调系统末端
air conditioned room 空气调节

A

房间

air conditioner 空调设备,空气调节器

air conditioning 空气调节

air conditioning area 空调面积

air conditioning equipment (ACE) 空气调节设备

air conditioning machine room 空调机房

air conditioning outlet 空调送风口

air conditioning plant 空调设备

air conditioning process (ACP) 空气调节过程

air conditioning system (ACS) 空气调节系统

air conditioning technique 空调技术

air conditioning theory 空气调节理论

air contaminant 空气污染物

air contamination 空气污染

air cooled air conditioner 风冷式空调器

air cooled condenser (ACC) 空气冷却冷凝器,空气冷凝器

air cooled condensing unit (ACCU) 风冷冷凝机组

air cooled cylinder 风冷气缸

air cooled type 风冷式

air cooler 冷风机

air cooling by evaporation 蒸发式空气冷却

air cooling fin 散热片

air curtain 空气幕

air cushion 空气垫

air damper 风阀,节气门,气流调节器

air defense 空防,防空

air dielectric coaxial 空气介质同轴(电缆)

air diffuser 散流器,空气扩散器

air distribution 风量分配,气流组织

air distribution equipment 空气分布设备

air distributor 空气分布器

air entraining concrete 加气混凝土

air escape valve 放气阀

air filter 空气过滤器

air filter apparatus 空气过滤装置

air filter regulator (AFR) 空气过滤调节器

air filter unit 空气过滤机组

air filtration 空气过滤

air flow window (AFW) 通风窗

air freezing 空气冻结

air freezing system 空气冻结系统

air friction 空气摩阻

air furnace 鼓风炉
air gauge 气压计
air grid 通风格栅
air grille 百叶[格栅]风口
air handler 空气处理机
air handling unit (AHU) 空气处理单元[机组],空调箱
air humidification 空气加湿
air humidifier 空气加湿器
air humidity 空气湿度
air induction 空气诱导
air infiltration 空气渗入
air inlet 空气入口,进风口
air inlet grille 进气格栅
air inlet valve 进气阀
air insulation 空气绝缘
air ion 空气离子
air jet 空气射流
air lance 空气枪
air leakage 漏气,空气渗漏
air liquefaction 空气液化
air lock 气塞[闸,孔]门斗,气锁阀
air meter 空气流量计
air moisture 空气含湿量
air motor 气压发动机,空气电机
air ozonizer 臭氧发生器
air parameter 空气参数
air passage 风道
air permeation 空气渗透
air pipe differential temperature fire detector 空气管差温火灾探测器
air pit 通风井
air pollutant 空气污染物
air pollution 空气污染
air pressure state 气压状态
air purifier 空气净化器
air quality 空气质量
air quality monitoring (AQM) 空气质量监测
air quality sensor (AQS) 空气质量传感器
air refrigerating machine 空气制冷机
air regulator 空气调节器,空气调节阀
air relief shaft 通风道
air renewal 换气
air return 回风
air shower 风淋室
air source heat pump system 空气源热泵系统
air supply grille 送风口
air to cloth ratio 气布比
air treatment 空气处理
air washer 空气洗涤器
air zoning 分区送风
air/fuel ratio control 空气燃料比控制

A

air-blast 鼓风,喷气

air-blast cooling (强制)通风冷却

air-blast freezer (强制)通风的冻结装置

air-blast freezing (强制)通风冻结

air-blast refrigeration 空气喷射制冷

airborne dust 气载[大气]尘埃

airborne particles 大气尘粒

airborne pollutant 风载污染物

air-conditioner for station 基站专用空调机

air-conditioning system 空调系统

airduct 通气道,风道

airflow 气流

airflow controller 空气流量调节器

airflow floor 通风地板

airflow meter 空气流量计

airflow resistance 气流阻力

airing priority 广播优先级

air-spaced coaxial cable 空隙同轴电缆

air-strainer 空气滤网

airstream 气流

air-supply mask 供气面罩

air-supported fiber (ASF) 空气间隙光纤

airtightness 气密,密封性

air-water system 空气-水系统

AIS (alarm indication signal) 告警指示信号

aisle 走道

AITS (acknowledged information transfer service) 确认信息传递服务

alarm (ALM) 告[报]警,警报

alarm and protection system 报警保护系统

alarm annunciator 警报信号器

alarm apparatus 报警器

alarm bell 火警警铃,警钟

alarm board 告警板

alarm box 告警箱

alarm bus 报警总线

alarm camera scanner 警报摄像机扫描器

alarm clock 报警时钟

alarm for oil-gas concentration 油气浓度报警器

alarm free 无报警

alarm indication signal (AIS) 告警指示信号

alarm indicator 告警指示器

alarm lamp 报警灯

alarm level 告警级别

alarm light 报警灯

alarm mainframe 报警主机

alarm management 报警管理

alarm message and alarm status

handling 告警提示处理

alarm module 告警模块

alarm of fire 火灾警报

alarm output 告警输出

alarm panel 告警面板

alarm point monitoring 报警点监控

alarm receipt 报警回执

alarm receiving center (ARC) 报警接收中心

alarm response time 报警响应时间

alarm signal 报警[警报]信号

alarm signal unit 报警信号单元

alarm software 报警软件

alarm source 告警源点

alarm subsystem 告警台

alarm system 报警系统

alarm whistle 警笛

alarming horn 报警喇叭

ALC (automatic level control) 自动电平控制

ALC control (automatic light compensation) 自动光量补偿

alcohol (乙)醇[酒精]

alcohol thermometer 酒精温度计

aldehyde (乙)醛

alertor 报警信号,报警器

algae control 藻类除去法

aliasing noise 混叠噪声

alien (exogenous) crosstalk 外部串扰

alien (exogenous) far-end crosstalk loss (AFEXT) 外部远端串扰损耗

alien (exogenous) near-end crosstalk loss (ANEXT) 外部近端串扰损耗

alignment 找平

alignment chart 诺模图

alkali cleaning 碱洗

all air heat recovery system 全空气热回收系统

all air system 全空气系统

all blast heating system 送风式供暖系统

all dielectric self-support (ADSS) 全介质自承式光缆

all optical network (AON) 全光网络

all purpose drying unit 通用干燥机

all water system 全水系统

all year air conditioning 全年空调

all-glass optical fiber 全玻璃光纤

allocation assignment 分配

allocator 分配器

allowable error 允许误差

allowable stress 允许应力

alloy steel 合金钢

A

all-plastic optical fiber (APOF) 全塑光纤

ALM (alarm) 告[报]警,警报

alphanumeric identifier 字母数字标识符

alternate layout 比较方案

alternating current (AC) 交流电

alternating current (AC) power conduit 交流电源管道

altimeter 高度计

altitude 标高,海拔高度

ALU/PETP-foil 覆塑铝箔

alumina 氧化铝,矾土

aluminum alloy wire (AAW) 铝合金线

aluminum foil 铝箔

aluminum foil shield 铝箔屏蔽

aluminum-alloy cable tray 铝合金电缆桥架

AM (amplitude modulation) 调幅

ambient air 环境[周围]空气

ambient environment 周围环境

ambient light 环境光

ambient noise 环境噪声

ambient temperature 环境温度

American National Standards Institute (ANSI) 美国国家标准协会

American National Standards Institute lumens (ANSI lumens) ANSI流明

American Society of Heating Refrigerating and Airconditioning Engineer (ASHRAE) 美国采暖制冷与空调工程师协会

American Telephone & Telegraph Inc. (AT & T) 美国电话电报公司

American wire gauge (AWG) 美国线规

amino group powder 氨基干粉

ammeter 电流[安培]表

ammonia compression refrigerator 氨压缩式制冷机

ammonia compressor 氨压缩机

ammonia condenser 氨冷凝器

ammonia cylinder 氨瓶

ammonia water 氨水

amount of heat 热量

AMP (AVGMaker Portable) 安普(一种综合布线系统品牌名)

AMP (amplifier) 放大器

amplified self-emission noise (ASE) 放大的自辐射噪声

amplifier (AMP) 放大器

amplitude modulation (AM) 调幅

amplitude-frequency response (AFR) 幅频响应

AMR (automatic meter reading system) 集中抄表系统

AMR (automatic meter reading)

自动抄表

AN (access network) 接入网

anaerobic digestion technology 厌氧消化技术

analog control 模拟控制

analog digital convertor 模拟数字[模数]转换器

analog input (AI) 模拟输入,模拟量输入

analog monitor 模拟监视器

analog output (AO) 模拟输出

analog signal 模拟信号

analog television (ATV) 模拟电视

analog video signal 模拟视频信号

analog video surveillance system 模拟视频监控系统

analog/digital (A/D) 模拟-数字

analog-to-digital (A/D) conversion 模数转换

analog-digital conversion (ADC) 模数转换

analogue controller 模拟控制器

analogue station set 模拟话机

analogue trunk unit (ATU) 模拟中继单元

analogy computer 模拟计算机

analysis layer 分析层

anchoring guy wire 锚拉线

ancillary data 辅助数据

anechoic room 无回声的房间[消声室]

anemometer 风速表

anemometrograph 风速风向记录仪,风速记录仪

anemoscope 风速计

ANEXT (alien/exogenous near-end crosstalk loss) 外部近端串扰损耗

angled outlet 斜口

angled patch panel 角型配线架

angled physical contact (APC) 成角度物理接触

angular sheet 角板

anion 阴离子

anion exchanger 阴离子交换器

anion generator 阴离子发生器

annealed copper 退火铜

annual cooling electricity consumption 空调年计算耗电量

annual energy efficiency ratio (AEER) 全年能效比

annual heating electricity consumption 采暖年计算耗电量

annular flow 环形流

anodized aluminum 阳极氧化铝

ANR (automatic network replenishment) 断网自动续传

ANS (automatic noise suppression) 自动噪声抑制

A

A

ANSI (American National Standards Institute) 美国国家标准协会
ANSI lumens (American National Standards Institute lumens) ANSI 流明
ante room 前室,准备室
antenna 天线
antenna amplifier 天线放大器
antenna and feeder 天线与馈线
antenna feed system 天线馈送系统
anti-icer 防冻装置
anti-passback (AP) 防反传
anti-rat bite 防鼠咬
anti-resonance 防共振[鸣]
anticoagulant 抗凝剂
anti-corrosion cable support system 防腐电缆桥架
anti-corrosive 防腐[蚀]剂
anti-dust cover 防尘盖
anti-dust cover for RJ45 plug RJ45 插头防尘盖
anti-fluctuator 稳压器
antifreeze agent 抗冻剂
anti-hunt action 抗振[阻尼]作用
antijamming 抗干扰
anti-kickback attachment 防反向安全装置
anti-rot 防腐
anti-rust 防锈
antiscale (锅炉)防垢剂
antiskid plate 防滑板
anti-skidding 防滑
anti-static wrist strap 防静电手腕带
anti-thrust bearing 止推轴承
AO (analog output) 模拟输出
AO (application outlet) 应用插座
AO (automation outlet) 自动化插座
AoC (advice of charge) 收费通知
AON (all optical network) 全光网络
AP (access point) (无线网的)接入点,访问点
AP (anti-passback) 防反传
apartment 公寓
APC (angled physical contact) 成角度物理接触
APD (avalanche photodiode) 雪崩光电二极管
aperture correction 孔径(失真)校正
API (application program interface) 应用程序接口
APL (acoustic power level) 声功率级
APOF (all-plastic optical fiber) 全塑光纤
APP (accident prevention program)

事故预防计划
apparatus attachment cord 装置连接跳线
apparatus dew point (ADP) 机器露点
apparent heat transfer coefficient 表面传热系数
apparent power 视在功率
appendage pump 备用泵
application of intrusion alarm system 入侵报警系统应用
application program interface (API) 应用程序接口
application server 应用程序服务器
application specific integrated circuit (ASIC) 专用集成电路
application temperature 应用温度
AQM (air quality monitoring) 空气质量监测
AQS (air quality sensor) 空气质量传感器
aqua 水柱(拉丁语,计量单位)
AR (augmented reality) 增强现实
aramid yarn 芳纶丝
ARC (alarm receiving centre) 报警接收中心
arc shape 弧面
arc welding electrode cable 电焊机电缆
architect 建筑师
architectural acoustics 建筑声学
archive 档案库
archive backup 归档备份
archive file 档案文件
area display 区域显示器
area heating 区域供暖
area number 区号
area of cooling surface 冷却面积
area of detection coverage 探测覆盖面
area of grate 炉算面积
area of safe operation 安全工作区
area of section 截面积
ARF (automatic roll filter) 自动卷绕式过滤器
arm stay 留守布防
armature 电枢
arming 布防
armor cash carrier 运钞车
array cabinet 列头柜
arrest point 临界点
arrester 制动器
arrow 箭头,指针
arson 放火(智能建筑设有纵火监控系统)
artificial atmosphere generator 人工气体发生器
artificial draft 人工通风
artificial ice 人造冰

artificial illumination 人工照明

artificial intelligence (AI) 人工智能

artificial light source 人工光源

artificial rainfall 人工降雨

artificial ventilation 人工通风

artisan 技工,工匠

as completed drawing 竣工图

ASA (acrylonitrile styrene acrylate copolymer) 丙烯腈-苯乙烯-丙烯酸

asbestos 石棉

as-built drawing 竣工图

ASC (automatic slope control) 自动斜率控制

ASE (amplified self-emission noise) 放大的自辐射噪声

ASF (air-supported fiber) 空气间隙光纤

ASHRAE (American Society of Heating Refrigerating and Airconditioning Engineer) 美国空调工程师协会

ASIC (application specific integrated circuit) 专用集成电路

ASIC and special chip ASIC 及专用电路芯片

ASL board with pulse billing 计费脉冲用户板

ASN.1 (abstract syntax notation one) 抽象描述语言(一种 ISO/ITU-T 标准)

ASON (automatically switched optical network) 自动交换光网络

aspect ratio 长宽比,宽高比

asphalt 沥青

aspherical 非球面的(镜片)

ASR (automatic speech recognition) 自动语音识别

assembly drawing 装配图

assembly drawing of the equipment 设备装配图

assembly hall 会堂

asset management 资产管理

asset management system 资产管理系统

assisted circulation boiler 辅助循环锅炉

Assmann aspiration psychrometer 阿斯曼干湿球湿度计

associated document 关联文档

asst project engineer 项目助理工程师

assurance factor 安全系数

assurance measure 保证措施

asymmetrical conductivity 非对称导电性

asymmetrical digital subscriber line (ADSL) 非对称数字用户线路

asymmetrical digital subscriber loop (ADSL) 非对称数字用户环路
asymmetrical element 非对称元件
asymmetrical modem 非对称调制解调器
asymmetrical modulation 非对称调制
asynchronous motor 异步电动机
asynchronous network 异步网络
asynchronous transfer 异步传输，异步切换
asynchronous transfer mode (ATM) 异步传输模式
asynchronous transmission (AT) 异步传输
AT (asynchronous transmission) 异步传输
AT & T (American Telephone & Telegraph Inc.) 美国电话电报公司
athermanous 不透热的
ATM (asynchronous transfer mode) 异步传输模式
ATM adaptation layer (AAL) 异步传输模式[ATM]适配层
ATM equipment 柜员机
atmometer 汽化[蒸发]计
atmospheric condenser 大气[淋浇]式冷凝器
atmospheric freeze drying (AFD) 常压冷冻干燥
atmospheric pressure 大气压力
atmospheric temperature 大气温度
atomization 雾化[喷雾]
atomizer 喷雾器
atomizing fineness 雾化细度
atomizing humidifier 喷雾加湿器
ATS (absolute temperature scale) 绝对温标
attached list 随箱清单
attachment screw 定位螺钉
attendance 员工考勤
attendant 运行[值班]人员，话务员
attenuation 衰减
attenuation coefficient 衰减系数
attenuation constant 衰减常数
attenuation to alien (exogenous) crosstalk ratio at the far-end (AACR-F) 远端衰减与外部串扰比
attenuation to alien (exogenous) crosstalk ratio at the near-end (AACR-N) 近端衰减与外部串扰比
attenuation to crosstalk ratio (ACR) 衰减串扰比
attenuation to crosstalk ratio at the far-end (ACR-F) 远端衰减与串扰比

attenuation to crosstalk ratio at the near-end (ACR-N) 近端衰减与串扰比

attenuator 衰减器

attenuator box 静压箱

attic ventilation 阁楼通风

attribute 属性

ATU (analogue trunk unit) 模拟中继单元

ATU-C (ADSL transmission unit-central) ADSL 传输中央单元(Modem)

ATU-R (ADSL transmission unit-remote) ADSL 远传单元(Modem)

ATV (analog television) 模拟电视

AU (administration unit) 管理单元

AU-AIS (administrative unit alarm indication signal) 管理单元告警指示信号

AU-LOP (loss of administrative unit pointer) 管理单元指针丢失

AU-PTR (administration unit pointer) 管理单元指针

AUC (authentication center) 鉴权中心

audio conference 音频会议

audible alarm 音响报警设备

audible signal 声音[音频]信号

audio & video control device 音像控制装置

audio band 音频频带

audio bandwidth 音频带宽

audio coding generation 3 (AC-3) 环绕声数字音频编码

audio compressor 音频压缩器

audio dub 音频配音

audio editing 音频编辑

audio equalization 音频均衡

audio exciter 音频激励器

audio frequency 音频频率

audio input/output 音频输入输出

audio matrix 音频矩阵

audio mode 音频模式

audio modem 音频调制解调器

audio noise 音频噪声

audio station 音频门口机

audio stream 音频流

audio video coding standard (AVS) 数字音视频编解码标准

audio video interleaving (AVI) 音视频交叉格式

audio workstation 音频工作站

audio/video (A/V) 音视频

audio-follow-video switcher 音频随视频切换器

audiometer 听力计

audio-video combiner 音频-视频组合器,视听传播合路器

A

audio-visual (AV) 音视频
audiovisual content 视音频内容，视听内容
audio-visual projector 音视频投影
audit trail 审计跟踪
auditorium 音乐厅
AUG (administrative unit group) 管理单元组
augmented reality (AR) 增强现实
AU-LOP (loss of administrative unit pointer) 管理单元指针丢失
authentication center (AUC) 鉴权中心
authentication information 鉴别信息
authentication, authorization, accounting (AAA) 认证,授权,计费
auto-back focus (ABF) 自动后焦调节
auto-iris lens 自动光圈镜头
auto-tuned control loop 自动调谐控制回路
auto-white balance (AWB) 自动白平衡
auto-alarm 自动报警
autocontrol 自动控制
autoconverter 自耦变压器
autogenous ignition temperature 自燃温度
automated infrastructure management (AIM) 自动化基础设施管理
automatic alarm 自动报警
automatic backlight compensation 自动背光补偿
automatic call distribution (ACD) 自动话务[呼叫]分配
automatic call distributor (ACD) 自动话务[呼叫]分配器
automatic control 自动控制
automatic detecting and recording system for violation of traffic signal 交通信号违章自动检测与记录系统
automatic electronic shutter (AES) 自动电子快门
automatic expansion valve (AEV) 自动膨胀阀
automatic explosion suppression system (AESS) 自动抑爆系统
automatic fire alarm and fire linkage system 火灾自动报警及消防联动系统
automatic fire alarm system 火灾自动报警系统
automatic fire detection system (AFDS) 自动火灾探测系统
automatic fire extinguisher 自动灭火装置
automatic fire signal 自动火灾

A

信号

automatic gain control (AGC) 自动增益控制

automatic level control (ALC) 自动电平控制

automatic light compensation (ALC) 自动光量补偿

automatic meter reading (AMR) 自动抄表

automatic meter reading system (AMR) 集中抄表系统

automatic network replenishment (ANR) 断网自动续传

automatic noise suppression (ANS) 自动噪声抑制

automatic protection 自动保护

automatic protection switching 自动保护切换

automatic protective system 自动保护系统

automatic roll filter (ARF) 自动卷绕式过滤器

automatic safety device 自动安全装置

automatic slope control (ASC) 自动斜率控制

automatic speech recognition (ASR) 自动语音识别

automatically switched optical network (ASON) 自动交换光网络

automation island 自动化岛

automation server 自动化服务器

automotive wire 汽车电线

AUX (auxiliary equipment) 辅助[附属]设备

auxiliary area 辅助区

auxiliary boiler 辅助锅炉(其锅炉房与周围建筑应按规定隔开相应距离)

auxiliary control element (ACE) 辅助控制单元

auxiliary data 辅助数据

auxiliary equipment (AUX) 辅助[附属]设备

auxiliary material 辅助材料

AV (audio-visual) 音视频

availability 可用性

avalanche photodiode (APD) 雪崩光电二极管

AVC (advanced video coding) 高级视频编码

average 平均数

average directional index (ADX) 平均趋向指数

average power sum alien (exogenous) near-end crosstalk loss 外部近端串扰损耗功率和平均值

average power sum alien near-end crosstalk (loss) (PS ANEXTavg) 外部近端串音功率和平均值

average power sum attenuation to alien (exogenous) crosstalk ratio far-end (PS AACR-Favg) 外部远端串扰损耗功率和平均值

average run power of single rack 单机架平均运行功率

AVI (audio video interleaving) 音视频交叉格式

avoiding lightning stroke 防雷

AVS (audio video coding standard) 数字音视频编解码(技术)标准

AWB (auto white balance) 自动白平衡

AWG (American wire gauge) 美国线规

axial fan 轴流风机

axial flow compressor 轴流式压缩机

axial velocity 轴向速度

axial-flow low noise fan 轴流低噪声风机

azeotrope 共沸混合物

azeotropic mixture 共沸混合物

azeotropic refrigerant 共沸制冷剂

B

B (baud)　波特(数据通信速度单位)

B channel (bearer channel)　B通道[信道],承载信道

B channel sequence number (BSN)　B通道顺序号

B frame　B帧

BA (BCCH allocation)　BCCH(频率)分配

BA (building automation)　楼宇自动化系统

baby bangor　小拉梯

BAC (building automation and control system)　建筑自动化和控制系统

back door　后门

back edge　后沿[下降边]

back filter　后置过滤器

back flame　复燃火焰

back focal distant (BFD)　后焦距

back focus　后焦点

back pack　背负式灭火器

back pack pump tank　背负式带泵灭火器

back pack pump tank fire extinguisher　小型背负泵式灭火器

back panel　后面板

back scattering　反[后]向散射

back stretch　反向铺设水带

back upright post　后立柱

back valve　单向阀,止回阀

back wiring　背面布线

backbone cable　主干线缆

backbone network　主干网(络)

back-end processor (BEP)　后端处理器

backfeed　反向馈电;反馈

backfill　回填

background broadcast　背景广播

background color　背景色

background music (BGM)　背景音乐

background music radio　背景音乐广播

background network　后台网络

backlight compensation (BLC)　背光补偿

backspace key 回退键
backtracking algorithm 回溯算法
back-up battery 备用电池
back-up breaker 备用断路器
back-up power 备用电源
back-up query 备份查询
back-up safety function 辅助安全功能
backward channel 反向信道
backward compatible 向后[下]兼容
backward curved impeller 后弯叶轮
backward search 反向搜索
backward supervision signal 后向监视信号
BACnet (a data communication protocol for building automation and control networks) 楼宇自动控制网络数据通信协议
BACnet (building automation and control networks) 楼宇自动化与控制网络
BACnet building controller (BBC) BACnet 楼宇控制器
BACnet IP Bacnet IP 协议
BACnet MS/TP BACnet 主从令牌数据链路协议
BACnet/IP BACnet 网关
bactericide 杀菌剂
bad sector 坏扇区
baffle 挡[隔]板，折流板
baffle evaporator 折流蒸发器
bag filter 布袋过滤器
bag type air filter 布袋滤尘器
balance flow 均衡流
balance pipe 均压管
balance tank 平衡罐
balance valve 平衡阀
balance ventilation 平衡通风
balanced cable 平衡[对绞]电缆，双绞电缆
balanced circuit 平衡电路
balanced cord 平衡跳线
balanced element 平衡元素
balanced network 对称网络，平衡网络
balanced protection 均衡防护
balanced signal pair 平衡线对
balanced system 均衡系统
balancing network 平衡网络
balancing pressure on stopping 均压防灭火
ball diffuser 球形送风口
ball float level controller 浮球式液位控制器
ball float valve 浮球阀
ball tube pulverizer 球管粉碎机，球磨机
band pass filter (BPF) 带通滤波器

B

bandwidth 频带宽度
bandwidth granularity 带宽间隔
bandwidth management 带宽管理
banking 工作的中止
BAOC (barring of all outgoing calls) 禁止所有呼出呼叫，闭锁全部去话
bar 巴(气压单位)，条，棒
bar code labelling 标识条码
barcode image 码图
barcode image version 码图版本
bare copper wire 裸铜丝
baroceptor 气压传感器
barograph 自记气压计
barrier frequency 截止频率
barring of all outgoing calls (BAOC) 禁止所有呼出呼叫，闭锁全部去话
BAS (building automation system) 楼宇自动化系统
base frame of cabinet 机柜底座
base injection foam extinguishing system 液下喷射[喷吹]泡沫灭火系统
base lighting 基础光
base requirement 基本需求
base station (BS) 基站
base station control function (BCF) 基站控制功能
base station controller (BSC) 基站控制器
base station identification (BSID) 基站识别码
base station identification information 基站识别信息
base station identity code (BSIC) 基站小区识别码
base station information 基站信息
base station manager 基站管理器
base station system (BSS) 基站子系统
base transceiver station (BTS) 基站收发信台，移动通信基站
baseband 基带
baseband coaxial cable 基带同轴电缆
baseband modem 基带调制解调器
basement 地下室
basic frame 基本框架
basic management grid 基本管理网格
basic rate ISDN 基本速率的综合业务数字网
basic service 基本业务
BAT (bouquet association table) 业务群关联表
battery 电池，蓄电池
battery capacity 电池容量
battery charger 蓄电池充电器
battery management system (BMS)

电池管理系统

battery room 电池室

baud (B) 波特(数据通信速度单位)

baud rate 波特率

bay window 凸窗

bayonet fibre optic connector (BFOC) 卡口式光纤连接器

BBAR (broad-band anti-reflective) 宽频率抗反射

BBAR multi-coating BBAR 多层镀膜(腾龙技术)

BBC (BACnet building controller) BACnet 楼宇控制器

BCCH (broadcast control channel) 广播控制信道

BCCH allocation (BA) BCCH(频率)分配

BCD (binary coded decimal) 二进制编码/十进制数

BCF (base station control function) 基站控制功能

BCST (broadcasting module) 广播模块

BCT (broadcast and communications technologies) 广播和通信技术

BD (building distributor) 建筑物配线设备,建筑物(主)配线架,楼宇配线架

beacon frequency 定标频率

beam 光束,电波

beam area 梁间区域

beam forming 波束成形

beam reception gain 受光增益

bearer channel (B channel) B 通道[信道],承载信道

BEF (building entrance facility) (建筑物)入口设施

BEL (bell character) 报警字符

bell 警铃

bell character (BEL) 报警字符

Bell Labs 贝尔通信实验室

bellows 波纹管

bend insensitive 弯曲优化,弯曲不敏感,抗弯

bend loss optimized (BLO) 弯曲损耗优化

bending loss 弯曲损耗

bending radius 弯曲半径

BEP (back-end processor) 后端处理机

BER (bit error rate) 信息差错率,比特误码率

BFD (back focal distant) 后焦距

BFOC (bayonet fibre optic connector) 卡口式光纤连接器

BG (border gateway) 边界网关

BGM (background music) 背景音乐

BGP (border gateway protocol) 边界网关协议

B

BI (binary input) 二进制输入
BICSI (Building Industry Consulting Service International) 国际建筑业咨询服务
bidder 投标人
bidding evaluation 评标
bidding specification 投标规格
bidirectional cable television transmission system 双向电缆电视传输系统
big space 大空间
billing system 计费系统
bin 二进制文件名后缀
bin string 二元符号串
binary coded decimal (BCD) 二进制编码/十进制数
binary digit bit 二进制数字位
binary input (BI) 二进制输入
binary phase shift keying (BPSK) 二进制相移键控
binary system 二进制
binder mechanism 进程间通信机制
binding 绑定
binding board 扎线板
biological assay 生物鉴定
biological attack 生物攻击
biological chip 生物芯片
biometrics 生物识别技术
BIP (bit interleaved parity) 位交织奇偶校验
BIS (building intelligent system) 建筑智能化系统
B-ISDN (broadband integrated services digital network) 宽带综合业务数字网
bit error rate (BER) 比特差错率
bit insertion 位插入
bit interleaved parity (BIP) 位交织奇偶校验
bit rate accuracy 比特率准确度
bit rate error tolerance 比特率容差
bit stream type 码流类型
bit string 位串
bit timing 位定时
bit transmission rate 位传输速率
bitrate stream smoothing 码流平滑
bitstream 位流
bitstream buffer 位流缓冲区
bitstream order 位流顺序
BLA (blocking-acknowledgement signal) 阻塞确认信号
black box testing approach 黑箱测试法
black burst 黑脉冲
black level 黑电平
black list (of IMEI) (IMEI 的)黑表

blank baffle 空白挡板
blank panel 空白面板
blanking retrace period 消隐折回周期
BLC (backlight compensation) 背光补偿
blemish 疵点
blind control 遮光帘控制
blind spot 盲点
blind test 盲测
blind zone 盲区
BLO (bend loss optimized) 弯曲损耗优化
block 闭塞,块
block scan 块扫描
blocking-acknowledgement signal (BLA) 阻塞确认信号
blooming 图像开花
blower extinguishment 风机灭火
blow-off valve seat 放水阀座
blue key 蓝色键
bluetooth 蓝牙
bluetooth wireless commissioning interface 蓝牙无线调试接口
blurring 图像模糊
BMS (battery management system) 电池管理系统
BMS (building management system) 建筑管理系统,建筑设备管理系统
BMW standard BMW(宝马)规格
BN (bonding network) 联结网络
BN (bridge number) 网桥号
BNC connector BNC 连接器
BNC female connector BNC 同轴插拔头(阴)
BNC male connector BNC 同轴插拔头(阳)
BNC terminator BNC 终接器
BNC-T connector BNC-T型接头
BO (broadcast outlet) 广播插座
board card 板卡
board name plate 板名条
board position 板位
boiler 锅炉
boiler room 锅炉房
boiler safety valve 锅炉安全阀
boil-over oil 沸溢性油品
bond 黏合剂
bonding 联结
bonding bar 等电位联结带
bonding conductor 等电位联结导体
bonding network (BN) 联结网络
boolean 开关量
booted soft wire 护套软线
border gateway (BG) 边界网关
border gateway protocol (BGP) 边界网关协议
BOSS (business & operation support

B

system) 业务运营支撑系统
bottom plate 底盘
boundary clock 边界时钟
boundary condition 边界条件
boundary value analysis (BVA) 边界值分析
bouquet 业务群
bouquet association table (BAT) 业务群关联表
bow-type drop cable 蝶形引入光缆
bow-type optical fibre cable 蝶形光缆
BPDU (bridge protocol data unit) 网桥协议数据单元
BPF (band pass filter) 带通滤波器
BPSK (binary phase shift keying) 二进制相移键控
bracket (墙上凸出的)托[支]架
braid 编织屏蔽
braid and foil screen 编织物和涂箔屏蔽
braid screen 编织物屏蔽
braided hose 编织软管
braided sleeving 编织套管
brain screen 编织屏蔽层
branched cable 分支线缆
BRAS (broadband remote access server) 宽带远程接入服务器
breach 攻破
break contact 动断触点
break point 断点
break time 分断时间
breakdown 崩溃,击穿
breakdown current 击穿电流
breakdown junction 击穿结
breakdown voltage 击穿电压
break-through 穿通
bridge connection 桥式连接
bridge converter 桥式转换器
bridge fault 桥接故障
bridge forwarding 网桥转发
bridge identifier 网桥标识符
bridge label 桥标记
bridge number (BN) 网桥号
bridge port 网桥端口
bridge protocol data unit (BPDU) 网桥协议数据单元
bridge static filtering (BSF) 网桥静态过滤
bridge tap 桥接分接头
bridging amplifier 桥接放大器
brightness 亮度
bring your own device (BYOD) 自带设备办公
Britain Standard (BS) 英国国家标准
broad-band anti-reflective (BBAR) 宽频率抗反射

broadband integrated services digital network (B-ISDN) 宽带综合业务数字网

broadband remote access server (BRAS) 宽带远程接入服务器

broadcast 广播

broadcast address 广播地址

broadcast and communications technologies (BCT) 广播和通信技术

broadcast audio extension 广播音频扩展

broadcast call 广播呼叫

broadcast control channel (BCCH) 广播控制信道

broadcast partition 广播分区

broadcast service 广播业务

broadcast television 广播电视

broadcasting application 广播方式应用

broadcasting module (BCST) 广播模块

broken link 断链

brush strip 刷条

BS (base station) 基站

BS (Britain Standard) 英国国家标准

BSC (base station controller) 基站控制器

BSF (bridge static filtering) 网桥静态过滤

BSIC (base station identity code) 基站小区识别码

BSID (base station identification) 基站识别码

BSN (B channel sequence number) B 通道顺序号

BSS (base station sub-system) 基站子[分]系统

BSS (business support system) 业务支撑系统

BSS operation and maintenance application part (BSSOMAP) BSS 运维应用部分

BSSOMAP (BSS operation and maintenance application part) BBSS 运维应用部分

BTS (base transceiver station) 基站收发信台,移动通信基站

BTS address management (BTSM) BTS 的站址管理

BTSM (BTS address management) BTS 的站址管理

BTU Meter 热量计

budget of installation 安装预算

BUF (buffer) 缓冲存储(器)

buffer (BUF) 缓冲存储(器)

builder 建筑者

building automation (BA) 楼宇自动化系统

B

building automation and control networks (BACnet) 楼宇自动化与控制网络
building automation and control system (BAC) 建筑自动化和控制系统
building automation system (BAS) 楼宇自动化系统
building backbone cable 建筑物主干缆线,楼宇主干线缆
building backbone cabling subsystem 干线子系统,楼宇主干布缆子系统
building block 积木式组件,积木
building column 楼宇柱体
building designer 建筑物设计者
building distributor (BD) 建筑物配线设备
building entrance facility (BEF) (建筑物)入口设施
Building Industry Consulting Service International (BICSI) 国际建筑业咨询服务
building information 建筑信息
building information system 建筑信息系统
building intelligent system (BIS) 建筑智能化系统
building intercom system 楼宇对讲系统
building management system (BMS) 建筑管理系统,建筑设备管理系统
building manager 建筑物经理
building regulation 建筑法规
building service 建筑设施
building structure 建筑结构
building structures sensor 建筑结构传感器
building unit 建筑单体
built-in fitting 预埋件
bulk range 喷射距离
bulk resistance 体电阻
bulkhead 隔板,隔离装置
bunched cable 集束电缆
bundled coaxial cables 集束同轴电缆
buoyant antenna 浮漂天线
burglar alarm 防盗报警器
burglar alarm system 防盗报警系统
burglar signal 安全防范信号
burglar-proof door 防盗门
burn-in 老化
burning behaviour 燃烧性能
burning material 燃烧材料
bus 总线,母线,汇流条,总线配置,导线
busbar 母线槽
business & operation support system (BOSS) 业务运营支撑系统

business announcement 业务广播

business support system (BSS) 业务支撑系统

butt joint 对接接头

butterfly valve 蝶阀

butting transmission system 对接传输设备

button 按钮

button switch 按键开关

butt-through 对穿

buzzer 蜂鸣器

BV 聚氯乙烯绝缘铜芯电线

BVA (boundary value analysis) 边界值分析

BYOD (bring your own device) 自带设备办公

bypass 旁路

bypass (access point) 门禁点旁路

bypass (zone) 防区旁路

bypass mode of UPS operation UPS的旁路运行方式

bypass power 旁路电源

bypass valve 旁通阀

byte alignment 字节对齐

byte number representation 字节的位号表示

C

C (connect) 连接

C/S (client and server) 客户端和服务器

C2C (consumer to consumer) 用户对用户的模式

CA (cell allocation) 小区[单元]分配

CA (communication automation) 通信自动化

CA (conditional access) 条件接收

cabinet 机柜

cabinet base frame 机柜底座

cabinet grounding wire 机柜接地导线

cabinet group 柜组

cabinet inter-connection hole 机柜互连孔

cable 电缆,线缆

cable assembly 电缆组件

cable box 电缆箱

cable bridge 电缆桥架

cable bundle 线[光]缆束

cable bundle jacket 光缆束护套

cable bus 电缆总线

cable carrier 托线架

cable chart 电缆(连接)图

cable cleat 电缆夹

cable communication 电缆通信

cable conduit of chlorinated polyvinyl chloride 氯化聚氯乙烯塑料电缆导管

cable conduit of fiber-cement 纤维水泥电缆导管

cable conduit of fiberglass reinforced plastic (FRP) 玻璃纤维增强塑料电缆导管

cable conduit of unplasticized polyvinyl chloride 硬聚氯乙烯塑料电缆导管

cable conduits of modified polypropylene 改性聚丙烯塑料电缆导管

cable connector 电缆连接器

cable cut-off wavelength (CCW) 光纤截止波长

cable design 电缆设计

cable digital television (CDT) 有线

数字电视

cable distribution equipment 电缆分线设备

cable distribution system 有线分配系统

cable distributor 电缆分支器

cable duct 电缆管道

cable element 电缆元素

cable equalization 电缆均衡

cable equalizer 电缆均衡器

cable fault 电缆故障

cable fault indicator (CFI) 电缆故障指示器

cable fill 电缆占用率

cable gland 电缆密封套

cable hut 电缆分线箱

cable installation work 电缆安装工程

cable interface unit (CIU) 电缆接口部件

cable labeling 电缆标识

cable ladder system 电缆梯架系统

cable laying 电缆敷设

cable layout 电缆布放

cable leak locator 电缆查漏仪

cable lug 电缆接线头

cable maintenance 电缆维护

cable management system 电缆管理系统

cable manhole 电缆人井

cable map 电缆图

cable modem (CM) 电缆调制解调器

cable modem termination system (CMTS) 电缆调制解调器终端系统

cable network cabinet 有线网集线间

cable outer diameter 电缆外径

cable rack 电缆支架

cable sandwich 电缆夹层

cable shaft 电缆竖井

cable sheath 电缆护套

cable shelf 电缆托架

cable shield 电缆屏蔽

cable stress 电缆应力

cable stripping tool 电缆剥线工具

cable system 电缆系统

cable telephony 有线电话(业务)

cable television (CATV) 有线电视

cable television network (CTN) 有线电视网

cable terminal 电缆终端

cable terminal box 电缆终端箱

cable terminator 电缆终端器

cable testing bridge 电缆故障测验桥

cable thermal detector (CTD) 缆式线型感温探测器

cable tie 电缆扎带

C

cable tray 电缆桥架
cable tray system 电缆托盘系统
cable trough 理线槽
cable trunking and ducting system 电缆槽管系统
cable tunnel 电缆隧道
cable type linear temperature fire detector 缆式线型感温火灾探测器
cable unit 电缆单元
cable vault 电缆(进线)室
cable-equalizing amplifier 电缆均衡放大器
cable-fault detector 电缆故障检验器
cable head 电缆分线盒
cable-joint protector box 电缆接头盒,电缆接头保护箱
cable-tray temperature sensor 缆式温度传感器
cabling 布线[缆]
cabling cabinet 配线机柜
cabling component 布线[缆]组件
cabling design document 布线设计文档
cabling diagram 电缆线路图
cabling screens 布缆屏蔽体
cabling subsystem 1 第一级子系统
cabling system for building 综合布线系统
CAC (channel access code) 信道访问码
cache duration 缓存时间
CAD (computer-aided design) 计算机辅助设计
CAD (computer-aided diagnosis) 计算机辅助诊断
CAE (computer-aided engineering) 计算机辅助工程
CAI (common air interface) 通用空中接口
CAI (computer-aided instruction) 计算机辅助教学
calamity detection 灾害探测
calibrate 校准
call accounting 呼叫计次
call accounting system (CAS) 呼叫计次系统
call address 呼叫[调入]地址
call control (CC) 呼叫控制
call control function (CCF) 呼叫控制功能
call information system (CIS) 呼叫信息系统
call station 寻呼台站
call-accepted condition 接收呼叫状态
call-accepted packet 呼叫接收包
call-accepted signal 接收呼叫信号

callin conference 呼入[被叫]型会议
calling channel 呼叫通道
callout conference 呼出[主叫]型会议
calorific potential 潜热能,发热量
CAM (computer-aided manufacturing) 计算机辅助制造
CAMA (centralized automatic message accounting) 集中式报文自动计费
camcorder 摄录一体机
camera 摄像机
camera control unit (CCU) 摄像机控制器
cameras and surveillance 视频监控
campus 建筑群,园区
campus backbone cable 建筑群[园区]主干缆线
campus backbone cabling subsystem 建筑群[园区]主干布缆子系统
campus distributor (CD) 建筑群[园区]配线设备
campus subsystem 建筑群[园区]子系统
CAN (controller area network) 控制器局域网络
canvas connecting terminal 帆布接头
capacitance 电容,电容量
capacitive displacement transducer (CDT) 电容式位移传感器
capacitor 电容器
capacitor type smoke detector 电容式感烟火灾探测器
capacity 容量
capillary 毛细管
capture ratio of vehicles 车辆(违章)捕获率
CAPWAP (control and provisioning of wireless access points protocol specification) 无线网接入点的控制和配置协议
carbon dioxide extinguishing agent 二氧化碳灭火剂
carbon dioxide extinguishing system 二氧化碳灭火系统
carbon dioxide fire extinguisher 二氧化碳灭火器
carbon monoxide canister 一氧化碳滤毒罐
carbon monoxide detector 一氧化碳探测器
card reader 读卡器
card reader/encoder 读卡器/编码器
cardholder 持卡人
carrier to interference ratio (CIR) 载波干扰比(载干比)
carrier to intermodulation ratio 载

波互调比

carrier to noise ratio (CNR) 载波噪声比

CAS (call accounting system) 呼叫计次系统

CAS (communication automation system) 通信自动化系统

CAS (conditional access system) 条件接收系统

cascading style sheet (CSS) 层叠样式表,样式级联表

CASE (computer-aided software engineering) 计算机辅助软件工程

CASS (conditional access subsystem) 条件接收子系统

casting-state structure 铸态组织

casualty 事故

CAT (conditional access table) 条件接收表

Cat.1 ～ Cat.8 综合布线产品传输分类：一类至八类

Cat.3 multi-pair cable 三类大对数电缆

Cat.6 cable 六类电缆

Cat.6 module 六类模块

Cat.6 non-shielded product 六类非屏蔽产品

Cat.6 non-shielded system 六类非屏蔽系统

Cat.6 RJ45 information module 六类 RJ45 型信息模块

Cat.6 stranded data jumper 六类数据多股跳线

Category 5 screened 五类屏蔽

Category 5 unscreened 五类非屏蔽

Category 6 screened 六类屏蔽

Category 6 unscreened 六类非屏蔽

category A1 multimode fibre A1 类多模光纤

catenary wire 吊索

cathode ray tube (CRT) 阴极射线管(显示器)

CATV (cable television) 有线电视

CATV (community antenna television or cable television) 共用天线(或有线)电视

CATV network 有线电视网络

CAVE (commputer automatic virtual environment) 计算机自动虚拟环境

CB (cell broadcast) 小区广播

CBN (common bonding network) 公共联结网络

CBR (constant bit rate) 固定码率,恒定比特率

CBR transport stream 恒定码率传送流

CBX (computerized branch exchange) 计算机化小交换机

CC (call control) 呼叫控制

CCA (cell coverage area) 小区覆盖范围

CCA (clear channel assessment) 空闲信道评估

CCA (cold consumption in unit air conditioning area) 单位空调面积耗冷量

CCAS (control center alarm system) 控制中心报警系统

CCAW (copper-clad aluminum wire) 铜包铝线

CCC (current carrying capacity) 载流量

CCCH (common control channel) 公用控制信道

CCD (charge-coupled device) 电荷耦合器件

CCF (call control function) 呼叫控制功能

CCF (cluster control function) 簇控制功能

CCIR (International Radio Consultative Committee) 国际无线电咨询委员会

CCIR-R.M.S. (Root Mean Square) 均方根有限值

CCITT (Consultative Committee of International Telegraph and Telephone) 国际电报电话咨询委员会

CCMS (central control and monitoring system) 中央监控系统

CCNTRL (centralized control) 集中控制

CCR (central control room) 中央控制室

CCT (correlated color temperature) 相关的颜色温度[色温(度)]

CCTV (close circuit television) 闭路电视

CCU (camera control unit) 摄像机控制器

CCU (central control unit) 中央控制主机

CCW (cable cutoff wavelength) 光纤截止波长

CD (campus distributor) 建筑群配线设备

CD (compact disc) 光碟,光盘

CD Library 光碟库

CD Tower 光碟塔

CDDI (copper distributed data interface) 铜缆分布式数据接口

CDMA (code division multiple access) 码分多址

CDN (content delivery network) 内容传递网

CD-ROM (compact disc-read only

memory) 只读碟

CDS (combined distribution system) 组合分配系统

CDT (cable digital television) 有线数字电视

CDT (capacitive displacement transducer) 电容式位移传感器

CE (Communate Europpene) 欧盟安全认证

CEBus (consumer electronics bus) 客户电子总线,消费电子总线

CECS (China Association for Engineering Construction Standardization) 中国建筑标准化协会

ceiling 吊顶

ceiling material 天花材料

ceiling screen 挡烟垂壁

cell 小区

cell allocation (CA) 小区[单元]分配

cell broadcast (CB) 小区广播

cell coverage area (CCA) 小区覆盖范围

cell identifier (CI) 小区标识号

cell identity (CI) 小区识别

cell loss priority (CLP) 信元丢失优先权

cell phone 移动电话

cell selection 小区选择

cell site controller (CSC) 小区控制器

cell site function (CSF) 信元位置功能,单基站功能控制

cellular broadcast channel 小区广播信道

cellular engineering 小区工程

cellular insulation 发泡绝缘

cellular mobile radio telephone service 蜂窝式移动无线电话业务

cellular mobile telephone network 蜂窝式移动电话网

cellular mobile telephone system (CMTS) 蜂窝式移动电话系统

celsius degree 摄氏温度

center cable network equipment room 有线网接入机房,中心电缆网络机房

center telecom equipment room 中心电信机房

centigrade 摄氏度

centigrade scale 摄氏温度

centigrade temperature 摄氏温度

central air conditioning thermostat 中央空调(风机盘管)温控器

central clock system 中央时钟系统

central control and monitoring system (CCMS) 中央监控系统

central control management equipment 中央控制管理设备

central control room (CCR) 中央控制室

central control unit (CCU) 中央控制主机

central fire alarm control unit 集中火灾报警控制器

central heating 集中供热

central management server (CMS) 监控中心系统,中央管理服务器

central processing unit (CPU) 中央处理器

central room for telecommunications 中央电信机房

central ventilation system (CVS) 中央通风系统

centralized alarm system 集中报警系统

centralized automatic message accounting (CAMA) 集中式报文自动计费

centralized billing 集中计费

centralized control (CCNTRL) 集中控制

centralized control type 集中控制型

centralized exchange function 集中用户交换机功能

centralized management 集中管理

centralized monitoring 集中监控

centralized monitoring system 集中监控系统

centralized payment 集中付费

centralized power non-centralized control type 集中电源非集中控制型

centralized supervisory and control equipment 集中监控设备

centralized topology 集中式拓扑结构

centre control unit 中央控制单元

centrifugal blower 离心鼓风机

centrifugal pump 离心泵

CEPT (Confederation of European Posts and Telecommunications) 欧洲邮电管理委员会

CER (common equipment room) 公共设备间

ceramic sleeve 陶瓷套管

certificate for first-class project manager 一级项目经理证书

certificate of conformity of product 产品(检验)合格证明

certificate of origin 原产地证明书,产地证

certificate of place of origin 原产地证明

CFI (cable fault indicator) 电缆故障指示器

C

CFM (**compounded frequency modulation**) 复合频率调制,复合调频

CGA (**color graphics adapter**) 色彩图形适配器

CGD (**combustible gas detector**) 可燃气体探测器

CGEM (**computable general equilibrium model**) 可计算一般均衡模型

CGI (**computer graphic interface**) 计算机图形接口

chairman unit 主席单元

challenge handshake authentication protocol (**CHAP**) 询问交接验证协议

change 变更

changed sound 变调声

changing device 变调器

channel 通道,信道,波道

channel access code (**CAC**) 信道访问码

channel analog process repeater 信道模拟处理中继器

channel digital process repeater 信道数字处理中继器

CHAP (**challenge handshake authentication protocol**) 询问交接验证协议

character generator 字符发生器

characteristic impedance 特性阻抗

charge (**CHG**) 负荷,电荷,充电,计费,收费

charge accounting 计费结算

charge index (**CHX**) 计费索引

charge mode 计费方式

charge rate (**CHR**) 计费费率

charge time zone 计费时区

charge-coupled device (**CCD**) 电荷耦合器件

charging device 计费设备

charging gateway 计费网关

charging subsystem (**CHS**) 计费子系统

chassis 机箱(计算机)

CHE (**compact heat exchanger**) 紧凑式换热器

check before acceptance 验收

chemical foam 化学泡沫

chemical reaction fire extinguisher 化学反应式灭火器

CHG (**charge**) 负荷,电荷,充电,计费,收费

chilled water cool 冷水式机组,冷冻水冷却

chiller 冷水机,冷机

chiller pump (**CHP**) 冰水泵

chiller unit 制冷机组

chimney effect 烟囱效应

China Association for Engineering Construction Standardization (CECS) 中国工程建设标准化协会
China mobile peer to peer (CMPP) 中国移动点对点协议
China State Bureau of Technical Supervision (CSBTS) 中国国家技术监督局
chip-set 芯片集
CHP (chiller pump) 冰水泵
CHR (charge rate) 计费费率
CHR (chiller return) 冷冻[冰]水回水端
chroma 色[浓]度
chroma corrector 色度校正器
chroma key 色度键
chroma noise 色度噪声
chromaticity 色度
chrominance level 色度电平
CHS (charging subsystem) 计费子系统
CHX (charge index) 计费索引
CI (cell identifier) 小区标识号
CI (cell identity) 小区识别
CI (common interface) 通用接口
CI (community intelligent) 住宅(小区)智能化
CIF (common intermediate format) 通用中间格式,CIF影像格式
CIFS (common internet file system) 公共网际文件系统(协议)
CIR (carrier to interference ratio) 载波干扰比(载干比)
circuit 电[回]路,线路
circuit breaker 断路器
circuit design 电路设计
circuit diagram 电路图
circuit group monitor message (GRM) 电路群监视消息
circuit integrity 线路完整性
circuit management 线路管理
circuit switched 电路交换
circuit switching exchange 电路交换的交换局
circuit-switched public data network (CSPDN) 电路交换公共数据网
CIS (call information system) 呼叫信息系统
CISC (complex instruction system computer) 复杂指令系统计算机
CISPR (International Special Committee on Radio Interference) 国际无线电干扰特别委员会
city data center of building energy consumption 城市建筑能耗数据中心
city grid management information system 城市网格化管理信息系统

C

CIU (cable interface unit) 电缆接口部件

civil engineering 土木工程

civil engineering structure 土木建筑结构

civilized construction 文明施工

cladding 包层

cladding of fiber 光纤包层

claim for compensation 索赔

clarification answering 澄清答疑

class 1 area 一类地区

class 2 area 二类地区

class 3 area 三类地区

Class C ～ Class FA 综合布线系统线路传输分级：C 级至 FA 级

classified management of plant 设备分级管理

CLB (current-limiting breaker) 限流断路器

clean class 洁净度

clean room 洁净室，无尘室

cleaning material 清洁液

cleaning rod 清洁棒

clear channel assessment (CCA) 空闲信道评估

client and server (C/S) 客户端和服务器

CLNP (connectionless network protocol) 无连接模式网络层协议

clock jitter 时钟抖动

close circuit television (CCTV) 闭路电视

closed device 封闭装置

closed length interval 封闭长度区间

closed trunking 封闭式线槽

closure for optical fiber cable 光缆接头盒

cloud computing 云计算

cloud security 云安全

cloud serving 云服务

cloud splicing 云拼接

cloud storage 云存储

cloud terrace 云台

CLP (cell loss priority) 信元丢失优先权

CLR (current-limiting resistor) 限流电阻器

cluster 簇，集群，群集，聚类

cluster control function (CCF) 簇控制功能

cluster switch system (CSS) 集群交换机系统

CM (cable modem) 电缆调制解调器

CM (configuration management) 配置管理

CM (cross-connect matrix) 交叉连接矩阵

C

CMI (coded mark inversion) 传号反转码

CMIS (common management information service) 公用管理信息服务

CMMS (computerized maintenance management system) 计算机维修管理系统

CMOS (complementary metal-oxide semiconductor) 互补金属氧化物半导体

CMP (communications plenum cable) 填实[干线]级通信电缆

CMPP (China mobile peer to peer) 中国移动点对点[对等]协议

CMS (central management server) 监控中心系统，中央管理服务器

CMS (configuration management system) 配置管理系统

CMTS (cable modem termination system) 电缆调制解调器终端系统

CMTS (cellular mobile telephone system) 蜂窝移动电话系统

CMX 住宅通信布线电缆

CN (core network) 核心网

CNA (computer network attack) 计算机网络攻击

CNC (computer numerical control) 计算机数控

CNR (carrier to noise ratio) 载波噪声比

CNS (control network system) 控制网络系统

coarse thread screw 粗牙螺丝

coarse thread screw hole 粗牙螺丝孔

coating 涂敷层

coax coupler 同轴电缆连接器

coaxial cable 同轴电缆

coaxial cable connector 同轴电缆连接器

coaxial communication cable 同轴通信电缆

coaxial pair 同轴线对

code division multiple access (CDMA) 码分多址

code for acceptance of construction quality 施工质量验收规范

code of acceptance of construction quality of electrical installation in building 建筑电气工程施工质量验收规范

CODEC (coder-decoder) 编解码器

coded mark inversion (CMI) 传号反转码

coded picture 编码图像

coder-decoder (CODEC) 编解码器

coding block 编码块

coding in colour television 彩色电

视编码
coding unit 编码单元
coefficient of performance (COP) 冷水机组运行效率，制冷性能系数
coefficient of safety 安全系数
coefficient of effective heat emission 有效热发射系数
cold consumption in unit air conditioning area (CCA) 单位空调面积耗冷量
cold test 低温试验
cold water system 冷却水系统
cold-pressed sheet steel 冷轧钢板
collecting bar 汇流条
collection layer 采集层，集合层
collision prevention 避碰
color balance 色彩平衡
color bar 色带[条]
color bar signal 彩条信号
color bit 彩色位
color burst 色同步信号
color coding 颜色编码
color contamination 串色
color correction 彩色校正
color decoder 色彩解码器
color depth 色深
color difference component 色差分量
color field 色彩场
color frame 色彩帧
color fringing 彩色镶边
color gamut 色域[阶]
color graphics adapter (CGA) 色彩图形适配器
color label 颜色标签
color signal 彩色信号
color space 色彩空间，色度空间
color subcarrier 色彩副载波
color temperature 色温
color-rendering properties of light source 光源的显色性
colour video signal 彩色视频信号
combination 组合
combination detector 复合探测器
combination type fire detector 复合式火灾探测器
combined agent extinguishing system 混合灭火系统
combined distribution system (CDS) 组合分配系统
combined front room 合用前室
combiner 合路器
combining transmission 并机发射
combustible gas alarm controller 可燃气体报警控制器
combustible gas detector (CGD) 可燃气体探测器
combustible material 可燃物
combustible vapor 可燃蒸气

combustion 燃烧

command and communication fire vehicle 通信指挥消防车

commissioning 试运行

commissioning period 试运行期

common air interface (CAI) 通用空中接口

common bonding network (CBN) 公共联结网络

common bus 公用总线

common bus system 公共总线系统

common bus topology 公共总线拓扑

common control channel (CCCH) 公用控制信道

common equipment room (CER) 公共设备间

common ground 共用接地

common HUB 普通型集线器

common infrastructure 共用基础设施

common interface (CI) 通用接口

common intermediate format (CIF) 通用中间格式,CIF影像格式

common internet file system (CIFS) 公共网际文件系统(协议)

common management information service (CMIS) 公用管理信息服务

common mode 共模

common security technology of information system 信息系统通用安全技术

common telecommunications room (CTR) 公共电信间

common-to-differential mode 共模至差模

commputer automatic virtual environment (CAVE) 计算机自动虚拟环境

Communate Europpene (CE) 欧盟安全认证

communication automation (CA) 通信自动化

communication automation system (CAS) 通信自动化系统

communication cable 通信电缆

communication module 通信模块

communication pipe well 通信管道井

communication room 通信机房

communication supporting installation 通信配套设施

communications hub 通信枢纽

communications plenum cable (CMP) 填实[干线]级通信电缆

community antenna television or cable television (CATV) 共用天线(或有线)电视

C

community intelligent (CI) 住宅(小区)智能化

community security 社区安防

compact disc (CD) 光碟

compact disc-read only memory (CD-ROM) 只读碟

compact heat exchanger (CHE) 紧凑式换热器

companion specification for energy metering (COSEM) 能源计量配套规范

comparative humidity 相对湿度

compatibility 兼[相]容性

compatible downward 向下兼容

compensated sample 补偿后样本

compressor 压缩机

complementary metal-oxide semiconductor (CMOS) 互补金属氧化物半导体

completion 竣工

completion drawing 竣工图纸

completion material 竣工资料

completion report 竣工报告

complex instruction system computer (CISC) 复杂指令系统计算机

compliance testing 符合性测试

component 构件,分量

component elementary stream 基本流分量

component standard 元器件标准

components in a colour television system 彩色电视系统中的分量

composite coaxial cable for telecommunication use 同轴综合通信电缆

composite video blanking and signal (CVBS) 复合视频基带信号

compounded frequency modulation (CFM) 复合频率调制,复合调频

compressed air 压缩空气

compression 压缩

compression ratio 压缩比

compression refrigerating machine 压缩式制冷机

compression stage 压缩级

compression/expansion of a picture 图像压缩扩展

computable general equilibrium model (CGEM) 可计算一般均衡模型

computer-aided design (CAD) 计算机辅助设计

computer-aided diagnosis (CAD) 计算机辅助诊断

computer-aided engineering (CAE) 计算机辅助工程

computer-aided manufacturing (CAM) 计算机辅助制造

computer-aided software engineering (CASE) 计算机辅助软件工程
computer communication 计算机通信
computer console 计算机控制台
computer graphic interface (CGI) 计算机图形接口
computer information system 计算机信息系统
computer network 计算机网络
computer network attack (CNA) 计算机网络攻击
computer network room 计算机网络机房
computer numerical control (CNC) 计算机数控
computer platform 计算机平台
computer power module 计算机专用电源
computer room 主机房
computer supported cooperative work (CSCW) 计算机支持的协同工作
computer telecommunication integration (CTI) 计算机电信集成
computer-aided instruction (CAI) 计算机辅助教学
computer-based language laboratory 计算机型语言实验室
computerized branch exchange (CBX) 计算机化小交换机
computerized maintenance management systems (CMMS) 计算机维修管理系统
concealed installation 暗装
concealed laying 暗敷
concealed work 隐蔽工程
concentrated air conditioning 集中式空调
concentrated heating 集中供暖
concentricity 同心[轴]度
concentricity error 同心[轴]度误差
condensate 冷凝水
condensation of moisture 结露
condensation point 露点
condenser 冷凝器
condenser started motor 电容起动电动机
condenser water pump (CWP) 冷凝泵
condensing apparatus 冷凝器
condensing pressure regulating valve 冷凝压力调节阀
condensing unit capacity 总制冷量
conditional access (CA) 条件接收
conditional access decoder 条件接收解码器
conditional access sub-system (CASS)

条件接收子系统

conditional access system (CAS) 条件接收系统

conditional access table (CAT) 条件接收表

conditioned air 空调空气

conductance 导热率

conductive floor 防静电地板

conductivity 电导率

conductivity for heat 导热系数

conductor 导体

conduit 管道,导管

conduit fixing device 导管固定装置

conduit laying 管道敷设

conduit system 导管系统

cone bearing 锥形轴承

cone valve 锥形阀

Confederation of European Posts and Telecommunications (CEPT) 欧洲邮电管理委员会

configuration management (CM) 配置管理

configuration management system (CMS) 配置管理系统

congeal 凝固

congealable point 凝固[结冻]点

congealation 冻结,凝固

connect (C) 连接

connecting hardware 连接(硬)件,连接器(硬)件

connection 连接

connection in parallel 并联

connectionless network protocol (CLNP) 无连接模式网络层协议

connector for optical fibre 光纤连接器

connector socket 接线盒插座

console air conditioner 托架式空调器

consolidation point (CP) 集合[汇集]点

consolidation point box 集合[汇集]点安装箱

consolidation point connector 集合[汇集]点连接器

consolidation point cord 集合[汇集]点跳线

consolidation point link (CPL) 集合[汇集]点链路

consortium (投标)联合体

constant bit rate (CBR) 固定码率,恒定比特率

constant bit rate coded video 恒定码率编码视频

constant pressure compression 等压压缩

constant pressure valve (CPV) 恒压阀

constant temperature and humidity system 恒温恒湿系统
constant temperature fire detector 定温火灾探测器
constant volume variable temperature system 定风量变温系统
constantan 康铜[铜镍合金]
constellation scrambling 星座扰码
construction area 建筑面积
construction ensemble system 建筑群子系统
construction organization plan 施工组织设计
construction process 施工工艺
construction process document 施工工艺文件
construction safety plan 施工安全计划
Consultative Committee of International Telegraph and Telephone (CCITT) 国际电报电话咨询委员会
consumable 耗材
consumer electronics bus (CEBus) 客户电子总线,消费电子总线
consumer to consumer (C2C) 用户对用户的模式
consuming management system 消费管理系统
contact impedance 接点阻抗
contact resistance 接触电阻
contact type 接触式
contactless 非接触
container 容器
containment spray system 安全壳喷淋系统
contaminant 污染物
contaminating 污染
content delivery network (CDN) 内容传递网
content provider 内容提供者[商]
continual service 持续业务
continuity of load power 负载电力的连续性
continuous power 实际持续功率
continuous severely errored second (CSES) 连续严重误码秒
continuous variable slope delta modulation (CVSDM) 连续可变斜率增量调制
contrast control 对比度控制
contrast enhancement 对比增强
contrast ratio 对比度
contrast resolution 对比度分辨率
control and provisioning of wireless access points protocol specification (CAPWAP) 无线接入点的控制和配置协议
control bus 控制总线
control cable 控制电缆
control center alarm system (CCAS)

C

控制中心报警系统
control logic 控制逻辑
control loop 控制回路
control network system (CNS) 控制网络系统
control signal 控制信号
control signal to start & stop an automatic equipment 自动设备启停联动控制信号
control system 控制系统
control valve 控制阀
control valve group (CVG) 控制阀组
control word (CW) 控制字
controller area network (CAN) 控制器局域网络
control program 控制程序
convection 对流
convectional cooling 对流冷却
convergence sublayer (CS) 会聚子层
conversion device 转换装置
converter 混频器,转换器
converter resolution 转换器分辨率
cooling capacity 制冷量
cooling tower 冷却水塔
cooling water 冷却水
cooperation 合作,配合
coordinated universal time (UTC) 协调通用(世界)标准时间
coordination 协调
coordinator 协调器
COP (coefficient of performance) 冷水机组运行效率,制冷性能系数
copper braid 铜丝网,铜编织物
copper clad aluminum power wire 铜包铝电源线
copper distributed data interface (CDDI) 铜缆分布式数据接口
copper patch cord 铜跳线
copper tape 铜带
copper-clad aluminum wire (CCAW) 铜包铝线
copper-conductor extruded plastic insulated cable with pre-fabricated 铜芯塑料绝缘预制分支电缆
copyright 著作权,版权
cord 跳线
cordless telephony system (CTS) 无线电话系统
cordon 警戒线
core diameter 纤芯直径
core network (CN) 核心网
corner 拐角
corona 电晕
correction factor 修正系数
correlated color temperature (CCT) 相关色温

correspond interruption rate 通信中断率
correspondence 对应
corresponding time 响应时间
corrosion-proof 防(腐)蚀
corrosive element 腐蚀元素
corrosiveness 腐蚀性
corrugated flexible metallic hose 波纹金属软管
corrugated steel tape (CST) 皱纹钢带
COSEM (companion specification for energy metering) 能源计量配套规范
country and city toll call area code table 国家和城市长途区号表
country code 国家码,国家代码
couple 耦合器
coupling 耦合
coupling of orbit and attitude 轨道和姿态耦合
coverage area 覆盖区
coverage rate 覆盖率
CP (consolidation point) 集合[汇集]点
CP cable 集合[汇集]点线缆
CP link (CPL) 集合[汇集]点链路
CPEV (constant pressure expansion valve) 定压膨胀阀
CPL (CP link) 集合[汇集]点链路
CPU (central processing unit) 中央处理器
CPV (constant pressure valve) 恒压阀
crack 裂纹
CRC (cyclic redundancy check) 循环冗余校验
creepage distance 爬电距离
crinoline 电缆导向装置
critical accident alarm 临界事故报警器
cross frame 十字骨架
cross modulation ratio 交扰调制比
cross section area (CSA) 横截面积
cross talk 串扰[音]
cross view 交叉视图,串画[像]
cross-connect 交叉连接
cross-connect matrix (CM) 交叉连接矩阵
cross-connecting distribution 交叉连接配线
cross-linked polyethylene insulated control cable 交联聚乙烯绝缘控制电缆
cross-linked polyolefin insulated wire 交联聚烯烃绝缘电线
cross-linked polyvinyl chloride insulated wire 交联聚氯乙烯绝缘电线
crossover frequency 分频点

cross-sectional area (CSA) 截面积
CRT (cathode ray tube) 阴极射线管(显示器)
crystal filter 晶体滤波器
crystal oscillator (XTLO) 晶体振荡器,晶体谐振器
CS (convergence sublayer) 会聚子层
CSA (cross section area) 横截面积
CSA (cross-sectional area) 截面积
CSBTS (China State Bureau of Technical Supervision) 中国国家技术监督局
CSC (cell site controller) 小区控制器
CSCW (computer supported cooperative work) 计算机支持的协同工作
CSES (continuous severely errored second) 连续严重误码秒
CSF (cell site function) 单基站功能控制,信元位置功能
CSI (current source inverter) 电流(源)型逆变器
CSPDN (circuit-switched public data network) 电路交换公共数据网
CSS (cascading style sheet) 层叠样式表,样式级联表
CSS (cluster switch system) 集群交换机系统
CST (corrugated steel tape) 皱纹钢带
CT (current transformer) 比流器,电流互感器
CTD (cable thermal detector) 缆式线型感温探测器
CTI (computer telecommunications integration) 计算机电信集成
CTN (cable television network) 有线电视网
CTR (common telecommunications room) 公共电信间
CTS (cordless telephony system) 无线电话系统
current carrying capacity (CCC) 载流量
current density 电流密度
current divider 分流器
current limit (control) 限流
current limiter 限流器
current limiting 限流
current loop 电流环
current loop interface 电流环接口
current source 电源
current source inverter (CSI) 电流(源)型逆变器
current transformer (CT) 比流器,电流互感器
current-limiting breaker (CLB) 限流断路器
current-limiting resistor (CLR) 限

流电阻器
cursor 光标
customs duty 关税
cut-off voltage 截止电压
cut-off waveguide vent 截止波导通风窗
cut-off wavelength 截止波长
cutout 保险开关，断路器
cut-over 割接
CVBS（composite video blanking and signal） 复合视频基带信号
CVG（control valve group） 控制阀组
CVS（central ventilation system） 中央通风系统
CVSDM（continuous variable slope delta modulation） 连续可变斜率增量调制
CW（control word） 控制字
CWP（condenser water pump） 冷凝泵
cycle life 循环寿命
cyclic redundancy check（CRC） 循环冗余校验

D

D

D/A (digital/analog) 数字/模拟

D1 D1 图像格式

DAB (digital audio broadcasting) 数字音频广播

DAC (discretionary access control) 自主访问控制

daisy-chain structure 菊花链结构

DALI (digital addressable lighting interface) 数字可寻址照明接口

DAM (digital audio machine) 数字音频播放器

damage accident 损坏事故

damaged length 烧毁长度

damper position 风门位置

damp-proof 防潮

dangerous accident 险性事故

DAP (digital audio player) 数字音频播放器

DAP (digital audio processor) 数字音频处理器

DARS (digital audio reference signal) 数字音频参考信号

DAS (direct-attached storage) 直接存储

DAS (distributed antenna system) 分布式天线系统

DAT (digital audio tape) 数字音频磁带

DAT (digital audio tape recorder) 数字磁带录音机

data acquiring subsystem 能耗数据采集子系统

data broadcasting 数据广播

data cable 数据电缆

data center (DC) 数据中心

data circuit terminating equipment (DCE) 数据电路端接设备

data communication network (DCN) 数据通信网(络)

data link control layer 数据链路控制层

data management server (DMS) 数据管理服务器

data module distribution frame 数据模块配线架

data of definite time and distance 定时定距数据

data of vehicle in and out parking lot or stop 车辆进出场(站)数据

data outlet 数据插座

data over cable service interface specification (DOCSIS) 有线电视数据业务接口规格

data point 数据点

data processing subsystem (DPS) 能耗数据处理子系统

data rate 码率

data receiving unit (DRU) 数据接收单元

data terminal (DT) 数据终端

data terminal equipment (DTE) 数据终端设备

data traffic of communication equipment 通信设备信息流量

data transfer rate (DTR) 数据传输(速)率

data transmission (DT) 数据传输

data transmission channel 数据传输通道

data transmission efficiency (DTE) 数据传输效率

data transmission interface 数据传输接口

data transmission ratio 数据传输比

data transmission speed 数据传输速度

data transmission system (DTS) 数据传输系统

data transmitting subsystem 数据传输子系统

data unit 数据单元

database (DB) 数据库

database management system (DBMS) 数据库管理系统

day template 日期模板

days of heating period 采暖期天数

DBC (direct buried cable) 直埋线缆

DBS (direct broadcasting satellite) 直播卫星

DBT (dry bulb temperature) 干球温度

DBX 压缩扩展式降噪系统

DC (data center) 数据中心

DC (decoder) 解码器

DC energy storage system 直流储能系统

DC link 直流环节

DC loop resistance 直流环路电阻

DC powered embedded thermal control equipment 直流供电嵌入式温控设备

DC resistance 直流电阻

DC resistance unbalance 直流电阻不平衡

D

DC-DC converter DC-DC变换器，直流-直流变流器

DC-dump 直流断电状态

DCE (data circuit terminating equipment) 数据电路端接设备

DCF (dispersion compensating fiber) 色散补偿光纤

DCIF (double CIF) DCIF图像格式

DCM (digital content manager) 数字内容管理器

DCM (door control module) 门禁控制模块

DCN (data communication network) 数据通信网(络)

DCP (digital cinema package) 数字电影包

DCR (drop call rate) 掉话率

DCS (distributed control system) 分散控制系统

DDC system (direct digital control system) 直接数字控制系统

DDE (dynamic data exchange) 动态数据交换

DDN (digital data network) 数字数据网(络)

DDNS (dynamic domain name server) 动态域名服务

DDoS (distributed deny-of-service) 分布式拒绝服务

DDR (digital disk recorder) 数字磁碟录像机

deadbeat control 无差拍控制

decentralized 分散式

decibels full scale 满度分贝

decimation filter 抽取滤波器

decision support system (DSS) 决策支持系统

decode 解码

decode time stamp (DTS) 解码时间戳

decoded picture 解码图像

decoded picture buffer (DPB) 解码图像缓冲区

decoder (DC) 解码器

decoder in colour television 彩色电视解码器

decoding 解码

decoding order 解码顺序

decorated wall 装饰墙

decoration system project 装饰系统工程

decryption 解密，译码

dedicated detection equipment 专用检测设备

dedicated ground 专用接地

dedicated line 专用线路

dedicated telephone 专用电话

default 预设，默认(值)

definition 分辨率，清晰度

defrost 除霜
degree of protection 防护等级
degree of safety 安全度
dehumidifying cooling 减湿冷却
delay correction 时延修正
delay deviation 延迟偏差
delay radio 迟播
delay skew 时延偏差
delay time 延迟时间
delivery multimedia integration framework (DMIF) 多媒体传送整体框架
delivery system 传送系统
deluge system 雨淋灭火系统
deluge valve group 雨淋阀组
demilitarized zone (DMZ) 隔离区
demultiplexer 多路分解器
deny override 超级解除
depth of field (DOF) 景深
dequantization 反量化
descrambler 解扰器
descrambling 解扰
descriptive cataloguing or indexing 描述式编目(或索引)
design 设计
design defect 设计缺陷
design development 深化设计
design document 设计文件
design drawing 设计图纸
design institute 设计单位
designer 设计者
detach 分离
detachable power supply cord 可拆卸的电源软线
detection 探[检]测
detection area 探测范围
detection area length 探测区域长度
detection distance 探测距离
detection perspective 探测视角
detection pressure 探测压力
detection response speed 探测响应速度
detection sensitivity 探测灵敏度
detection zone 探测区域
detector 探测器
deterioration 变质
device administrator 设备管理员
device description 设备描述(表)
device interface 设备接口
device link detection protocol (DLDP) 设备连接检测协议
dew point (DP) 露点
dew point temperature 露点温度
DFF (dispersion flattened fiber) 色散平坦光纤
DHCP (dynamic host configuration protocol) 动态主机配置协议
DI (digital input) 数字输入
dial 拨号[盘]

D

dial backup 拨号备份
dial exchange 拨号交换机
dial mode 拨号方式
dial modem 拨号调制解调器
dial network 拨号网(络)
dial number indentification service (DNIS) 拨号识别服务
dial phone set 拨盘式电话,号盘话机
dialing 拨号
dialing area 拨号区
dialing directory 拨号目录
diameter of conductor 导体直径
diameter over-insulated conductor 绝缘导体直径
dielectric constant 介电常数
dielectric strength 介电强度
dielectric test 绝缘试验
differential mode 差模
differential mode delay (DMD) 差模延迟
differential pressure switch (DPS) 差压开关
differential pulse code modulation (DPCM) 差分脉冲编码调制
differential temperature detector 差温探测器
differential-to-common mode 差模至共模
differentiated service (DiffServ) 区分服务
diffuser air supply 散流器
digital control 数字控制
digital addressable lighting interface (DALI) 数字可寻址照明接口
digital audio broadcasting (DAB) 数字音频广播
digital audio machine (DAM) 数字音频播放器
digital audio player (DAP) 数字音频播放器
digital audio processor (DAP) 数字音频处理器
digital audio reference signal (DARS) 数字音频参考信号
digital audio tape (DAT) 数字音频磁带
digital audio tape recorder (DAT) 数字磁带录音机
digital cinema package (DCP) 数字电影包
digital content manager (DCM) 数字内容管理器
digital cross connect (DXC) 数字交叉连接,数字交接设备
digital data network (DDN) 数字数据网
digital disk recorder (DDR) 数字磁碟录像机
digital gas meter 数字燃气表

digital hard disk recorder 数字硬盘录像机

digital heat meter 数字热量表

digital information system (DIS) 数字信息系统

digital input (DI) 数字输入

digital light processing (DLP) 数字光学处理

digital light processor (DLP) 数字光学处理器

digital micromirror device (DMD) 数字微镜元件

digital noise reduce (DNR) 数字降噪

digital output (DO) 数字输出

digital rights management (DRM) 数字版权管理

digital satellite system (DSS) 数字卫星系统

digital signage system 数字告示系统

digital signal processing (DSP) 数字信号处理

digital signature 数字签名

digital subscriber line (DSL) 数字用户线路

digital television (DTV) 数字电视

digital terrestrial television broadcasting repeater 地面数字电视广播中继站

digital theater system (DTS) 数字影院系统

digital transmitter communicator 数字传输通信机

digital video broadcast (DVB) 数字视频广播

digital video intercom 数字可视对讲

digital video recorder (DVR) 数字录像机

digital video server (DVS) 数字视频服务器

digital video surveillance system (DVSS) 数字视频监控系统

digital visual interface (DVI) 数字视频接口

digital visual interface-analog (DVI-A) 数字视频接口-模拟(DVI-A)

digital visual interface-digital (DVI-D) 数字视频接口-数字(DVI-D)

digital visual interface-integrated (DVI-I) 数字视频接口-集成

digital water meter 数字水表

digital zero 数字零

digital/analog (D/A) 数字-模拟

D-ILA (direct-drive image light amplifier) 直接驱动图像光源放大器(技术)

dimmer 调光器

dimming module 调光模块

DIP switch (DIP) DIP 开关

direct authorization 直接授权

direct broadcast 直播

direct broadcasting satellite (DBS) 直播卫星

direct broadcasting satellite designated service area 直播卫星服务区

direct burial 直埋

direct buried cable (DBC) 直埋电缆

direct connection 直接连接

direct contact 直接接触

direct current loop resistance 直流环路电阻

direct digital control system (DDC system) 直接数字控制系统

direct laying (电缆)直接敷设

direct outward dialing one (DODⅠ) 直接向外拨号一次,一次拨号音

direct return system (DRS) 异程式系统,直接回水系统

direct-attached storage (DAS) 直接存储

direct-connecting distribution 直接配线

direct-drive image light amplifier (D-ILA) 直接驱动图像光源放大器(技术)

direction of lay 绞合方向

directional response characteristics (DRC) 定向型响应特性

directional wheel 方向轮

directivity 指向性

directivity pattern 指向性模式

DIS (digital information system) 数字信息系统

disarm 撤防

disarm condition 解除状态

discrete frequency 离散频率

discretionary access control (DAC) 自主访问控制

dismountable type 可拆卸式

dispatch center 调度中心

dispatching 调度

dispersion compensating fiber (DCF) 色散补偿光纤

dispersion flattened fiber (DFF) 色散平坦光纤

dispersion phenomenon 色散现象

dispersion shifted fiber (DSF) 色散移位光纤

displacement factor 位移因数

displacement ventilation 置换通风

display multimedia 显示媒体

display order 显示顺序

display screen 显示屏

display screen fixed on roof 屋顶显示屏

display screen fixed on wall 墙面显示屏

disruption 中断

distortion 失真，畸变

distortion of long duration of a picture 长时间图像失真

distortion of short duration of a picture 短时间图像失真

distributed antenna system (DAS) 分布式天线系统

distributed building service 分布式楼宇设施

distributed control system (DCS) 分散控制系统

distributed deny-of-service (DDoS) 分布式拒绝服务

distributed linear optical fiber temperature fire detector 分布式线型光纤感温火灾探测器

distributed return loss (DRL) 分布式回波损耗

distributed structure 分布式结构

distribution equipment 配线设备

distribution frame 配线架

distribution management system (DMS) 配电管理系统

distribution network automation (DNA) 配电网自动化

distribution point (DP) 分配点

distribution room 配线间

distribution splitter 分支保护器

distributor 分配器，配线架

distributor 1 第一级配线设备

district cooling 区域供冷

diverse-routed protection (DRP) 多路由保护

diversity RF port 分集射频端口

diverting valve 分流阀

DLDP (device link detection protocol) 设备连接检测协议

DLP (digital light processing) 数字光处理

DLP (digital light processor) 数字光学处理器

DMD (differential mode delay) 差模延迟

DMD (digital micromirror device) 数字微镜元件

DMIF (delivery multimedia integration framework) 多媒体传送整体框架

DMS (distribution management system) 配电管理系统

DMZ (demilitarized zone) 隔离区

DNA (distribution network automation) 配网自动化

DNIS (dial number indentification service) 拨号识别服务

DNR (digital noise reduce) 数字降噪

D

DNS (domain name system) 域名系统

DO (digital output) 数字输出

DOC (documentation) 文档

DOCSIS (data over cable service interface specification) 有线电视数据业务接口规格

document object model (DOM) 文档对象模型

documentation (DOC) 文档

DODI (direct outward dialing one) 直接向外拨号一次,一次拨号音

DOF (depth of field) 景深

Dolby atmos 杜比全景声

Dolby E 杜比 E

Dolby noise reduction 杜比降噪

Dolby Pro Logic 杜比定向逻辑(环绕声)

DOM (document object model) 文档对象模型

domain name system (DNS) 域名系统

door contact 门磁(开关)

door control hardware 门控五金件

door control module (DCM) 门控模块

door control relay 门控继电器

door entry control 门禁

door status monitor (DSM) 门状态监视器

double arc shape 双弧面

double blind test 双盲测试

double CIF (DCIF) DCIF 图像格式

double glazing with shutter 中空百叶玻璃,双层百叶窗

double wall corrugated cable conduit of chlorinated polyvinyl chloride 氯化聚氯乙烯塑料双壁波纹电缆导管

double wall corrugated cable conduit of unplasticized polyvinyl chloride 硬聚氯乙烯塑料双壁波纹电缆导管

DP (dew point) 露点

DP (distribution point) 分配点

DP (dynamic programming) 动态规划

DPB (decoded picture buffer) 解码图像缓冲区

DPCM (differential pulse code modulation) 差分脉冲编码调制

DPS (data processing subsystem) 能耗数据处理子系统

DPS (differential pressure switch) 差压开关

drain wire 汇流线

drawer with telescopic slide 伸缩滑轨抽屉

DRC (directional response characteristics) 定向型响应特性
drencher system 水幕系统
DRL (distributed return loss) 分布式回波损耗
DRL0 分布式回波损耗常量
DRM (digital rights management) 数字版权管理
DRM agent DRM[数字版权管理]代理
DRM content DRM[数字版权管理]内容
DRM server DRM[数字版权管理]服务器
drop call rate (DCR) 掉话率
drop optical fibre cable 引入光缆
dropout 信号失落
DRP (diverse-routed protection) 多路由保护
DRS (direct return system) 异程式系统,直接回水系统
DRU (data receiving unit) 数据接收单元
dry and wet bulb thermometer 干湿球温度表
dry bulb temperature (DBT) 干球温度
dry coil 干式盘管
dry coil unit 干式盘管装置
dry contact digital input 干触点数字输入
dry heat test 高温试验
dry pipe system 干式系统
dry powder fire extinguishing system 干粉灭火系统
dry sprinkler system 干式喷水灭火系统
dry type transformer (DTT) 干式变压器
dry-chemical fire extinguisher 干粉灭火器
DSF (dispersion shifted fiber) 色散移位光纤
DSL (digital subscriber line) 数字用户线路
DSM (door status monitor) 门状态监视器
DSP (digital signal processing) 数字信号处理
DSS (decision support system) 决策支持系统
DSS (digital satellite system) 数字卫星系统
DT (data terminal) 数据终端
DT (data transmission) 数据传输
DTE (data terminal equipment) 数据终端设备
DTE (data transmission efficiency) 数据传输效率
DTMF (dual tone multi-frequency)

双音多频

DTR (data transfer rate) 数据传输(速)率

DTS (data transmission system) 数据传输系统

DTS (decode time stamp) 解码时间戳

DTS (digital theater system) 数字影院系统

DTT (dry type transformer) 干式变压器

DTV (digital television) 数字电视

dual cool 双冷源式

dual core LC coupler 双芯 LC 耦合器

dual core LC-LC multi-mode optical fiber jumper wire 双芯LC-LC多模光纤跳线

dual function system 双功能系统

dual port faceplate 双口面板

dual port oblique cutting wall faceplate 双口斜插墙面面板

dual tone multi-frequency (DTMF) 双音多频

duct mounted installation 管道安装

duplex 双重的,双工的,双向的

duplex adapter 双工适配器

duplex cable 双绞电缆

duplex clip 双口夹

duplex coupler 双工耦合器

duplex SC connector (SC-D) 双工SC 连接器

durability 耐久性

duration of fire resistance 耐火极限

duress 挟持,胁迫

dust 粉尘

dust protection 防尘保护

dust shutter 防尘盖

dust-proof cover 防尘罩

duty grid 责任网格

DVB (digital video broadcast) 数字视频广播

DVB-Cable (DVB-C) 有线数字视频广播

DVI (digital visual interface) 数字视频接口

DVI-A (digital visual interface-analog) 数字视频接口-模拟(DVI-A)

DVI-D (digital visual interface-digital) 数字视频接口-数字(DVI-D)

DVI-I (digital visual interface-integrated) 数字视频接口-集成

DVR (digital video recorder) 数字录像机

DVS (digital video server) 数字视频服务器

DVSS (digital video surveillance system) 数字视频监控系统

DXC (digital cross connect) 数字交叉连接

dynamic data exchange (DDE) 动态数据交换

dynamic domain name server (DDNS) 动态域名服务(器)

dynamic host configuration protocol (DHCP) 动态主机配置协议

E

E

E2E (end-to-end) 端对[到]端
early detection 早期探测
earth current 大地电流
earth electrode 接地体
earth path 接地路径
earth resistance 接地电阻
earthing 接地
earthing bus conductor 接地干线
earthing conductor 接地导体,接地线
earthquake 地震
EBM (emergency broadcasting message) 应急[紧急]广播信息
EC (echo canceler) 回声消除器
EC (encoder) 编码器
ECA (energy consumption in unit air conditioning area) 单位空调面积能耗
ECC (embedded control channel) 嵌入式控制通道
eccentricity 偏心率,偏心度
ECD (electro-chromic chemical display) 电化变色显示器,电致变色化学显示器
echo 回声,回执,回送
echo canceler (EC) 回声消除器
echo canceler pool (GECP) GSM 回波抵消板,回波消除器池
echo of videoconferencing 电视会议回声
ECM (entitlement control message) 授权控制信息
ECMA (European Computer Manufactures Association) 欧洲计算机制造联合会
ECMAScript ECMAScript 脚本语言
EDA (equipment distribution area) 设备配线区
edge blending 边缘融合
EDI (electronic data interchange) 电子数据交换
EDID (extended display identification data) 扩展显示标识数据
EDO (electric door opener) 电动开门器
EEAT (electric energy acquisition

terminal) 电能采集终端

EEG (energy efficiency grade) 能效等级

EER (energy efficiency ratio) 能效比

EERr (energy efficiency ratio of refrigeration system) 制冷系统能效比

EERs (energy efficiency ratio of air conditioning system) 空调系统能效比

EERt (energy efficiency ratio of terminal system) 终端系统能效比

effective acoustic center 有效声源中心

effective aperture 有效孔径

effective area utilization 有效面积利用率

effective bandwidth 有效带宽

effective spot 有效光斑

EG (electronic government) 电子政务

EGA (enhanced graphics adapter) 增强图形适配器

EI (equipment interface) 设备接口

eigen frequency 特征频率

EIS (electronic image stabilization) 电子稳像[防抖]

EIT (event information table) 事件信息表

EL (electro-luminescent) 电致发光

ELC (electronic light control) 电子光控

electric bell 电铃

electric cable 电缆

electric circuit 电路

electric connection 电连接

electric contact set 电接触装置

electric control 电气控制

electric door opener (EDO) 电动开门器

electric energy acquisition terminal (EEAT) 电能采集终端

electric exit device 电控逃生装置

electric field strength 电场强度

electric fireproof valve 电动防火阀

electric fuse 电熔丝

electric generator set 发电机组

electric lock 电子锁

electric lock controller 电子锁控制器

electric panel heating 电热板供暖

electric power line 电力线

electric power supply system (EPSS) 供(配)电系统

electric power supply system project

供(配)电系统工程
electric relay 电气继电器
electric rotating machine (ERM) 旋转电机
electric rule checking (ERC) 电气规则检查
electric specification 电气规范
electric strike 电控锁扣,电动扣板
electric throttle actuator 电动节流阀执行机构
electric valve 电动阀
electric wire 电线[缆]
electrical cable 电力电缆
electrical cable conduit 电缆管道
electrical fire 电气火灾
electrical fire alarm sounder 电气火灾警报器
electrical fire monitoring system 电气火灾监控系统
electrical line 电气线路
electrical machine 电机,发电机
electrical pipe well 电气管道井
electrical protection 电气防护
electrical safety 电气安全
electrical safety marking 电气安全标志
electrical safety measure 电气安全措施
electrical safety regulation 电气安全规程
electrical safety standard 电气安全标准
electrical safety test 电气安全试验
electrical schematic 电气原理图,电路图
electrical zero 电零位
electricity leakage protection device 漏电保护装置
electricity meter 电表
electrified locksets & trim 电控锁和电控把手
electro-chromic chemical display (ECD) 电化变色显示器,电致变色化学显示器
electro-control anti-burglary door 电控防盗门
electro-control lock 电控锁
electro-luminescent (EL) 电致发光
electro-magnetic compatibility (EMC) 电磁兼容性
electrode cable detector 电极电缆(故障)探测器
electromagnet 电磁铁
electromagnetic brake 电磁制动器
electromagnetic compatibility (EMC) 电磁兼容性
electromagnetic field 电磁场

E

electromagnetic flowmeter 电磁流量计

electromagnetic interference (EMI) 电磁干扰

electromagnetic protection 电磁保护

electromagnetic radiation 电磁辐射

electromagnetic screen 电磁屏蔽

electromagnetic shield 电磁屏蔽(物)

electromagnetic shielding 电磁屏蔽

electromagnetic shielding enclosure 电磁屏蔽室

electromagnetic valve 电磁阀

electromagnetic wave 电磁波

electromechanical device 机电装置

electromotive force (EMF) 电动势

electron noise 电子噪声

electronic (power) switch 电子(电力)开关

electronic ballast 电子镇流器

electronic color filter 电子滤色器

electronic component 电子元件

electronic data interchange (EDI) 电子数据交换

electronic electric energy meter 电子式电能计量装置

electronic equipment 电子设备

electronic equipment room 电子设备间

electronic equipment room system project 电子设备机房系统工程

electronic government (EG) 电子政务

electronic image stabilization (EIS) 电子稳像[防抖]

electronic information system 电子信息系统

electronic information system room 电子信息系统机房

electronic key 电子钥匙，电子密钥

electronic light control (ELC) 电子光控

electronic multi-functional electric energy meter 多功能电子电能表

electronic news gathering (ENG) 电子新闻采集

electronic patch panel (EPP) 电子配线架

electronic program guide (EPG) 电子节目指南

electronic sand table (EST) 电子沙盘

electronic serial number (ESN) 电子序列号

E

E

electronic signage 电子招牌

Electronic Testing Laboratories (ETL) 电子测试实验室

electronic voting 电子投票

electronic whiteboard 电子白板

electronic-mail (E-MAIL) 电子邮件

electrophoretic coating 电泳镀层

electrophoretic indication display 电泳显示

electrostatic 静电

electrostatic discharge 静电放电

electrostatic discharge earthing 静电放电接地

electrostatic harm 静电危害

electrostatic leakage 静电泄放

electrostatic shield 静电屏蔽

elementary examination 初验

elementary stream (ES) 基本(码)流

elementary stream clock reference (ESCR) 基本流时钟基准

elevated fire tank 高位消防水箱

elevated floor 架空地板

elevator cable 电梯电缆

elevator car 电梯轿厢

elevator controlling equipment 电梯控制器,梯控

elevator conversion layer 电梯转换层

elevator machine room (EMR) 电梯机房

elevator system 电梯系统

ELFEXT (equal level far-end crosstalk ratio) 等电平远端串扰衰减

ELTCTL (equal level TCTL) 两端等效横向转换损耗

ELV (extra-low voltage) 特低压

EM performance 电磁性能

E-MAIL (electronic-mail) 电子邮件

embedded control channel (ECC) 嵌入式控制通道

embedded multi-media card (eMMC) 嵌入式多媒体卡

embedded thermal control equipment 嵌入式温控设备

embedding material 灌封材料

EMC (electro-magnetic compatibility) 电磁兼容性

EMC directive 电磁兼容指令

EMCS (energy management and control system) 能源管控系统

emergency broadcast 应急[紧急]广播

emergency broadcasting 应急[紧急]广播设备

emergency broadcasting data segment 应急[紧急]广播数据段

emergency broadcasting message

(EBM) 应急[紧急]广播信息

emergency intercom 应急[紧急]内部通话系统

emergency lighting 应急[紧急]照明

emergency lighting centralized power supply 应急[紧急]照明集中电源

emergency lighting controller 应急[紧急]照明控制器

emergency lighting distribution equipment 应急[紧急]照明配电设备

emergency maintenance center 应急[紧急]维修中心

emergency optical fibre cable 应急[紧急]光缆

emergency plan 应急[紧急]预案

emergency power 应急[紧急]电源

emergency pulse 呼救脉冲

emergency response system 应急[紧急]响应系统

emergency safety procedure 应急[紧急]安全程序

emergency signal 应急[紧急]信号

emergency socket 应急[紧急]插座

emergency start signal 应急[紧急]启动信号

emergency switch 应急[紧急]开关

emergency telephone 应急[紧急]电话

EMF (electromotive force) 电动势

EMI (electromagnetic interference) 电磁干扰

EMM (entitlement management message) 授权管理信息

eMMC (embedded multi media card) 嵌入式多媒体卡

employee entrance 员工入口处

EMR (elevator machine room) 电梯机房

EMS (element management system) 网元[元素]管理系统

EMS (energy management system) 能源管理系统

EN (European Norm) 欧洲标准

enclosed staircase 封闭楼梯间

encoder (EC) 编码器

encoding presentation 编码表示

encrypt 加密

encryption 加密

end 末端

end-node of IoT 物联网端节点

end-point 终端,端点

end-to-end (E2E) 端对[到]端

end-user 端点[最终]用户

end-wall 端墙

end-device 终端设备

E

end-to-end link 端到端链路
end-user 终端用户
energy center 能源中心
energy consumption in unit air conditioning area (ECA) 单位空调面积能耗
energy consumption of different items 分项能耗
energy consumption of different sort 分类能耗
energy efficiency 能量效率
energy efficiency grade (EEG) 能效等级
energy efficiency ratio (EER) 能效比
energy efficiency ratio of air conditioning system (EERs) 空调系统能效比
energy efficiency ratio of air-conditioner 空调设备的能效比
energy efficiency ratio of refrigeration system (EERr) 制冷系统能效比
energy efficiency ratio of terminal system (EERt) 终端系统能效比
energy management 能源管理,能效管理
energy management and control system (EMCS) 能源管控系统
energy management system (EMS) 能源管理系统
energy metering 能源计量
energy metering application 能源计量应用
ENG (electronic news gathering) 电子新闻采集
engineering accident 工程事故
engineering architecture 工程建筑
engineering investigation 工程考察
engineering of security and protection system (ESPS) 安全防范(系统)工程
engineering qualitative accident 工程质量事故
engineering structural accident 工程结构事故
enhanced graphics adapter (EGA) 增强图形适配器
ENI (external network interface) 外部网络接口
EnOcean (wireless energy harvesting technology) 易能森无线能量采集技术
ensured function 应备功能
ensured sound level 应备声压级
entering air temp 进风温度
entering water temp 进水温度
enterprise fire station 企业消防站
enthalpy 焓(值)

enthalpy entropy chart 焓熵图
entire life 总寿命
entitle 授权
entitlement control message (ECM) 授权控制信息
entitlement management message (EMM) 授权管理信息
entrance and exit 出入口
entrance facility 入口设施
entrance machine 门口机
entrance pipe 进楼管
entrance point 入口点
entrance point from the exterior of the building 楼宇外部入口点
entrance room 进线间
entropy coding 熵编码
entry 项目,入口,条目,表目
entry/exit control 进出控制
environment of anti-electrostatic discharge 抗静电(放电)环境
environment system 环境系统
environmental control 环境控制
EO (equipment outlet) 设备插座
EPG (electronic program guide) 电子节目指南
epoch 历元
EPON (Ethernet passive optical network) 以太(网)无源光网络
EPP (electronic patch panel) 电子配线架
EPSS (electric power supply system) 供电系统
ePTZ 电子云台控制
EQ (equalizer) 均衡器
EQP (equipment) 设备
equal level far-end crosstalk ratio (ELFEXT) 等电平远端串扰衰减
equal level TCTL (ELTCTL) 两端等效横向转换损耗
equalizer (EQ) 均衡器
equipment (EQP) 设备
equipment cable 设备缆线
equipment controller 设备控制器
equipment cord 设备跳线
equipment distribution area (EDA) 设备配线区
equipment for building-in 嵌入式设备
equipment installation 设备安装
equipment interface (EI) 设备接口
equipment machine room 设备机房
equipment manager 设备管理器
equipment outlet (EO) 设备插座
equipment rack 设备机架
equipment room (ER) 设备用房,设备间
equipment sample 设备样品

E

E

equipment status monitoring (ESM) 设备状态监测

equipotential bonding 等电位联结

ER (equipment room) 设备用房，设备间

ERB (existing residential building) 既有住宅

ERC (electric rule checking) 电气规则检查

ERM (electric rotating machine) 旋转电机

ERPS (Ethernet ring protection switching) 以太网多环保护技术

error 差错，出错，误差

error concealment 错误隐藏

error correction level 纠错等级

error ratio 出错率，错误率

errored second (ES) 误码秒

errored second ratio (ESR) 误码秒比率

ES (elementary stream) 基本(码)流

ES (errored second) 误码秒

eSATA (external serial ATA) SATA 接口的外部扩展规范，外部串行 ATA

escape exit 紧急出口

escape hatch 逃生人孔，紧急出口

ESCR (elementary stream clock reference) 基本流时钟基准

ESD (electro-static discharge) 静电释放

ESD earth bonding point ESD 接大地连接点

ESM (equipment status monitoring) 设备状态监测

ESN (electronic serial number) 电子序列号

ESPS (engineering of security and protection system) 安全防范(系统)工程

ESR (errored second ratio) 误码秒比率

EST (electronic sand table) 电子沙盘

Ethernet 以太网

Ethernet passive optical network (EPON) 以太网无源光网络

Ethernet port 以太网端口

Ethernet ring protection switching (ERPS) 以太网多环保护交换技术

Ethernet switch 以太网交换机

ETL (Electronic Testing Laboratories) 电子测试实验室

EUI-64 (64-bit extended unique identifier) 64 位扩展的唯一标识符

European Computer Manufactures Association (ECMA) 欧洲计算机制造联合会
European Norm (EN) 欧洲标准
evacuated export 疏散出口
evacuation indication system 疏散指示系统
evacuation route 疏散通道
evacuation signal 疏散信号
evaluating value of energy conservation 节能评价值
evaporator 蒸发器
event information table (EIT) 事件信息表
event log 事件日志
exact match 精确匹配
excerpt 片段
excess cable length tray 冗余缆线存储盒,多余电缆长度托盘
excess current 过载电流
excess heat 余热
excessive floor loading 楼板超载
exciter 激励器
executive privileges 执行权限
executive regulation mechanism 执行调节机构
ex-factory permit 出厂许可证
ex-factory testing 出厂检验
ex-factory testing recording 出厂检验记录
ex-factory testing report 出厂检验报告
exhaust fan 排烟风机
exhaust hood 排风罩,排气罩
exhaust smoke port 排烟口
exhaust smoke window 排烟窗
existing residential building (ERB) 既有住宅
exit 出口
exit direction sign 出口方向标志
expansion plan 扩容方案
expansion tank 膨胀水箱
expected 预料
expedited forwarding 快速转发
expert control 专家控制
explicit declaration 显式声明
explicit expiration time 显式过期时间
explicitly undefined behavior 显式未定义行为
explosion-proof 防爆
exposed 明敷
extended display identification data (EDID) 扩展显示标识数据
extended EPG information 扩展EPG信息
extended station 分机
extensible hypertext markup language (XHTML) 可扩展超文本标记语言

E

extensible markup language (XML) 可扩展标记语言
extension 分机
extension cabinet 副[辅助]机柜
extension number 分机号
extension project 扩建工程
extension transit 分机转接
exterior wall 外墙
external cable 外部线缆
external cycle 外循环
external network 外部网络
external network interface (ENI) 外部网络接口
external serial ATA (eSATA) SATA 接口的外部扩展规范，外部串行 ATA
external stairway 室外楼梯
extinguishing agent 灭火剂
extractable drawer 可抽出抽屉
extra-low voltage (ELV) 特低电压
extra-low voltage circuit 特低压电路

F

F (flag)　标志

F (frequency)　频率

F frame　F 帧

F/FTP　金属箔线对屏蔽与金属箔总屏蔽对绞电缆

F/UTP　金属箔总屏蔽对绞电缆

F2TP　双层金属箔总屏蔽对绞电缆

FA (factory automation)　工厂自动化

FA (fire automation)　消防自动化

fabrication　制造

FAC (fire alarm controller)　火灾报警控制器

FACCH (fast-associated control channel)　快速随路控制信道

FACCH/F (fast-associated control channel/full rate)　全速率快速随路控制信道

face recognition from still image and video　基于静止图像和视频的人脸识别

faceplate　面板

facility　(基础)设施

facility code　工厂代码

facsimile apparatus　传真机

factory automation (FA)　工厂自动化

factory information system (FIS)　企业信息管理系统

factory inspection　工厂检验

factory manufacture supervision　工厂监造

FACU (fire alarm control unit)　火警报警控制装置

fade in　淡入

fade out　淡出

fading　衰减，淡入淡出

Fahrenheit scale　华氏温标

Fahrenheit temperature scale (°F)　华氏温度

FAI (fire alarm installation)　火灾报警装置

fail safe　掉电安全模式

failure　漏报

failure locating　故障定位

false alarm 误报警
false alarm of fire 假火警
fan box 风扇盒
fan coil 风机盘管
fan coil unit (FCU) 风机盘管
fan filter unit (FFU) 风机过滤网[机]组
fanout cable 扇出光缆
far-end block error (FEBE) 远端(成)块误码
far-end crosstalk (FEXT) 远端串扰
far-end crosstalk attenuation (loss) (FEXT) 远端串音[扰]衰减(损耗)
FAS (fire alarm system) 火灾报警系统
FAS (fresh air supply) 新风供给
fast-associated control channel (FACCH) 快速随路控制信道
fast-associated control channel/full rate (FACCH/F) 全速率快速随路控制信道
fast Ethernet 快速以太网
fast flashing 快闪
fast reverse playback 快速倒放
fastener 紧固件
fastening bolt 紧固螺栓
fastening screw 紧固螺钉
FAU (fresh air unit) 新风机组
fault 故障
fault alarm 故障报警
fault block 故障闭塞
fault complaint handling 故障申告受理
fault isolation 故障隔离
fault locating method (FLM) 故障定位方法
fault management 故障管理
fault signal 故障信号
fault status information 故障状态信息
fault time 故障时间
fault tolerant 容错
fault-tolerance assurance plan 容错保证方案
faulty convergence 错误收敛[会聚]
FB (feed back) 反馈
FBI (fiber interface) 光纤接口
FC (fiber channel) 光纤信道
FC (fiber connector) FC 光纤连接器
FC bus FC 总线
FC connector (full contact connector) FC 型光纤连接器
FCA (fixed channel allocation) 固定信道分配
FCA (fixed channel assignment) 固定信道布置

FCC (**Federal Communications Commission**) 美国联邦通信委员会

FCoE (**fiber channel over Ethernet**) 以太网光纤通道

FCR (**fire control room**) 消防控制室

FCS (**fieldbus control system**) 现场总线控制系统

FCS (**fire control system**) 灭火控制系统

FCS (**frame check sequence**) 帧校验序列

FCU (**fan coil unit**) 风机盘管

FD (**floor distributor**) 楼层配线架,楼层配线设备

FDB (**floor distributor box**) 楼层集线箱

FDDI (**fiber distributed data interface**) 光纤分布数据接口

FDMA (**frequency division multiple access**) 分频多址

FE (**functional entity**) 功能实体

feature 特征

FEBE (**far end block error**) 远端(成)块误码

Federal Communications Commission (**FCC**) 美国联邦通信委员会

feed 馈送

feed back (**FB**) 反馈

feed forward structure 前馈结构

feed voltage test 馈电电压测试

feedback 反馈

feedback exterminator 反馈抑制器

feedback signal from automatic equipment 自动设备反馈信号,联动反馈信号

feeder cable 馈线电缆

feeder connection 馈线连接

feeder fixing clip 馈管固定夹

feeder grounding clip 馈管接地夹

feeder in a cable distribution system 有线分配系统中的馈线

feeder window 馈线窗

feedthrough capacitor 穿心式电容器

fence 围栏

FEP (**fluorinated ethylene propylene**) 氟化乙烯丙烯共聚物

FES (**front end system**) 前端系统

FET (**field effect transistor**) 场效应管

FEXT (**far-end crosstalk**) 远端串扰

FEXT (**far-end crosstalk attenuation/loss**) 远端串音[扰]衰减(损耗)

FF (**foundation fieldbus**) 基础现场总线

FFC (**flat flexible cable**) 柔性扁平

带状电缆
FFC (flexible flat cable) 柔软带状电缆
FFS (for further study) 待研究
FFU (fan filter unit) 风机过滤[网]机组

FHS (fire hydrant system) 消火栓系统
fiber 纤维,光纤
fiber access network 光纤接入网(络)
fiber attenuator 光纤衰减器
fiber bandwidth 光纤带宽
fiber buffer 光纤缓冲层
fiber bundle 光纤束
fiber cable 光缆
fiber channel over Ethernet (FCoE) 以太网光纤通道
fiber conduit 光纤导管
fiber connector (FC) FC光纤连接器
fiber coupler 光纤耦合器
fiber cut 断纤
fiber distributed data interface (FDDI) 光纤分布数据接口
fiber grating 光纤光栅
fiber in the loop 光纤用户环路
fiber interface (FBI) 光纤接口
fiber linker 光纤连接器
fiber optic cable (FOC) 光缆
fiber optic connector (FOC) 光纤连接器
fiber optic coupler (FOC) 光纤耦合器
fiber optic drop 光纤入户线
fiber optic jumper 光纤跳线
fiber optic transceiver (FOT) 光纤收发器
fiber optics cluster 光导纤维束
fiber pigtail 光纤尾纤
fiber plus copper structure 光纤加铜芯结构
fiber splice tray 光纤熔接盒
fiber to the building (FTTB) 光纤到楼
fiber to the curb (FTTC) 光纤到路边
fiber to the desk (FTTD) 光纤到桌面
fiber to the floor (FTTF) 光纤到楼层
fiber to the home (FTTH) 光纤到户
fiber to the node (FTTN) 光纤到节点
fiber to the office (FTTO) 光纤到办公室
fiber to the premises (FTTP) 光纤到驻地
fiber to the remote (FTTR) 光纤

到远端

fiber to the service area (FSA) 光纤到服务区

fiber to the subscriber 光纤到用户单元

fiber to the subscriber unit communication facility 光纤到用户单元通信设施

fiber transmission 光纤传输

fiberglass protective layer 玻璃纤维保护层

fibre distribution box 光纤配线箱

fibre end-face 光纤端面

fibre optic adaptor type SC SC 型光纤适配器

fibre optic inter-repeater link (FOIRL) 光纤中继器间链路

fictitious load 假负载

field 域,场

field angle 视场角,光斑角

field bus 现场总线

field construction management agreement 现场施工管理协议

field effect transistor (FET) 场效应管

field inspection 现场考察

field test equipment 现场测试设备

field tester 现场测试仪

fieldbus 现场总线

fieldbus control system (FCS) 现场总线控制系统

field-mountable optical connector 现场组装式光纤(活动)连接器

field-programmable gate array (FPGA) 现场可编程门阵列

file transfer protocol (FTP) 文件传输协议

filter 过滤网,滤波器

filtered sample 滤波后样本

final acceptance 最终验收

finder 取景器

fine tuning 微调

finer granularity 细粒度

fingerprint identification 指纹采集仪

finished board manufactured board 成品板制成板

finite impulse response filter (FIRF) 有限冲激响应滤波器

fire 火,火焰,火灾

fire alarm control unit (FACU) 火警报警控制装置

fire alarm controller (FAC) 火灾报警控制器

fire alarm device 火灾警报装置

fire alarm information 火灾报警信息

fire alarm installation (FAI) 火灾报警装置

F

fire alarm system (FAS) 火灾报警系统
fire alarm transmission equipment 火灾报警传输设备
fire auto-alarm system 火灾自动报警系统
fire automation (FA) 消防自动化
fire broadcasting 消防广播
fire classification 火灾分类
Fire code 法尔码
fire communication equipment 消防通信设备
fire communication system 消防通信系统
fire compartment 防火分区
fire compartmentation 防火分隔
fire confirmation light 火警确认灯
fire control room (FCR) 消防控制室
fire control system (FCS) 灭火控制系统,消防系统
fire danger 火灾危险
fire detector 火灾探测器
fire display plate 火灾显示盘
fire door 防火门
fire door monitor 防火门监控器
fire duty room 消防值班室
fire elevator front room 消防电梯前室
fire elevator room 消防电梯机房
fire emergency broadcast 消防应急[紧急]广播
fire emergency broadcast control device 消防应急[紧急]广播控制装置
fire emergency broadcast system 消防应急[紧急]广播系统
fire emergency lighting 消防应急照明
fire facility 消防设施
fire hazard 火灾危害性
fire hydrant button 消火栓按钮
fire hydrant pump 消火栓泵
fire hydrant system (FHS) 消火栓系统
fire insulation 耐火隔热性
fire integrity 耐火完整性
fire lift 消防电梯
fire linkage controller 消防联动控制器
fire linkage equipment 消防联动设备
fire linkage system 消防联动系统
fire load 火灾荷载
fire load density (FLD) 火灾荷载密度
fire parameter 火灾参数
fire phone 消防电话
fire point 燃点
fire power monitor 消防电源监

控器

fire prevention 防火

fire propagation hazard 火焰传播危险

fire protection 消防

fire protection control room 消防控制室

fire protection system project 消防系统工程

fire public address (FPA) 火灾事故广播

fire public device (FPD) 消防公共设备

fire pump 消防泵

fire pump room 消防水泵房

fire resistance cables tray 耐火电缆槽盒

fire resistance classification 耐火等级

fire resistant cable 耐火电缆

fire resistant damper (FRD) 防火阀

fire resistant shutter (FRS) 防火卷帘

fire resisting beam 耐火梁

fire resisting column 耐火柱

fire resisting damper (FRD) 防火阀

fire resisting duct 耐火管道

fire resisting floor 耐火楼板

fire resisting partition 耐火隔墙

fire resisting roof 耐火屋顶

fire resisting suspended ceiling 耐火吊顶

fire resisting window 防火窗

fire resistive 耐火的

fire resistive cable 耐火电缆

fire retardance 阻燃性

fire retardant 阻燃剂

fire retardant treatment (FRT) 阻燃处理

fire risk 火灾危险性

fire scene 火灾场景[现场]

fire shutter controller 防火卷帘控制器

fire sound alarm 火灾声警报器

fire sound and light alarm 火灾声光警报器

fire special telephone 消防专用电话

fire special telephone extension 消防专用电话分机

fire special telephone switchboard 消防专用电话交换机

fire stability 耐火稳定性

fire suppression 灭火

fire telephone 火警电话

fire telephone line 火警电话线

fire transmission equipment 火警传输设备

fire wall 防火墙

fire zone 火灾区域

fireplug 消防栓

fireproof coating 防火涂料

fireproof coating for electric cable 电缆防火涂料

fireproof door 防火门

fireproof sealing 防火封堵

fireproof shutter 防火卷帘

fire-resistant cable trunking 耐火电缆槽盒

firewall 防火墙

firewire 火线

FIRF (finite impulse response filter) 有限冲激响应滤波器

first class low current system worker 一级弱电工

first class power supply 一类市电供电

first degree fault 一级故障

first-party release 第一方释放,第一方拆线

FIS (factory information system) 企业信息管理系统

fittings for commissioning 调测配件

five level low current system worker 五级弱电工

fix 固定

fixation kit 固定组件

fixed channel allocation (FCA) 固定信道分配

fixed channel assignment (FCA) 固定信道布置

fixed connector 固定连接器

fixed equipment 固定安装设备

fixed focal 定焦的

fixed focus 固定物距,定焦

fixed horizontal cable 固定水平线缆

fixed IT service 固定 IT 服务

fixed satellite service (FSS) 固定卫星业务

fixed station 固定台

fixed temperature detector 定温探测器

fixed wavelength filter 固定波长滤波器

fixed wireless access (FWA) 固定无线接入

fixed-mobile convergence (FMC) 固网移动融合

fixing glue 固定胶

fixture 夹具

flag (F) 标志

flame 火焰

flame detector 火焰探测器

flame front 火焰峰,火焰前缘

flame propagation 火焰传播

flame retardancy/low smoke zero

halogen (FR/LSOH) 阻燃低烟无卤

flame retardant cable 阻燃电缆

flame retardant refractory wire and cable 阻燃耐火电线电缆

flame spread 火焰传播

flame spread rate (FSR) 火焰传播速率

flame spread time (FST) 火焰传播时间

flame-inhibiting stabilizing element 防火稳定元素

flameless combustion 无焰燃烧

flame-resistant security cable 耐火安全电缆

flame-retardant 阻燃

flaming 有焰燃烧

flammable and explosive place 易燃易爆场所

flammable gas 可燃[易燃]气体

flash 闪燃

flat cable 扁平电缆

flat cable connector 带状电缆连接器

flat fiber optic cable 扁平光缆

flat flexible cable (FFC) 柔性扁平带状电缆

flat outlet 平口插座

FLD (fire load density) 火灾荷载密度

flexible (metal) electrical conduit 可挠性(金属)电气导管

flexible cable 柔性线缆,软电缆

flexible data cable 数据软线

flexible flat cable (FFC) 柔软带状电缆

flexible flat cable connector 柔软带状电缆连接器

flexible point (FP) (业务)灵活点

flexible rubber-sheathed cable 橡套软电缆

flicker 闪烁

flight simulator 飞行模拟器

FLM (fault locating method) 故障定位方法

float type 浮点型(数据)

floating charge 浮充

floating mode 浮动模式

floating nut 浮动螺母

flooding 扩散,泛洪

floor 楼板,楼层

floor box 地面插座[安装]盒

floor cable 楼层线缆

floor distributor (FD) 楼层配线架,楼层配线设备

floor distributor box (FDB) 楼层集线箱

floor distributor box with power 楼层有源集线箱

floor panel heating (FPH) 地板辐

射采暖

flow divider 分流阀

flow meter 流量计

flow monitoring 流量监控

flow switch 流量开关

flow velocity 流速

flowmeter 流量计

fluorinated ethylene propylene (FEP) 氟化乙烯丙烯共聚物

flush-mounted installation 嵌入安装,暗装

FMC (fixed-mobile convergence) 固网移动融合

FMD (follow-me diversion) 跟我转移

***F*-number** *f* 值,焦距比数

fiber optic adapter 光纤适配器

foam extinguishing device 泡沫灭火装置

foam extinguishing system 泡沫灭火系统

FOC (fiber optic cable) 光缆

FOC (fiber optic connector) 光纤连接器

FOC (fiber optic coupler) 光纤耦合器

focal length (镜头)焦距

focus 焦距

foil pair screen 线对箔屏蔽层

foil screen 金属箔屏蔽

foil twisted pair (FTP) 金属箔屏蔽双绞线电缆

FOIRL (fibre optic inter-repeater link) 光纤中继器间链路

folding frequency 折叠频率

follow-me diversion (FMD) 跟我转移

follow-on current 后续电流

for further study (FFS) 待研究

for reference 供参考

forbidden 禁止

force arm away 强制外出布防

force arm stay 强制留守布防

force arming 强制布防

forced cut 强制切入,强切

forced ventilation 强制通风

foreign associated construction project 涉外建设项目

forklift 叉车

form factor 波形因数

formality expense 手续费

format 格式,格式化

FOT (fiber optic transceiver) 光纤收发器

foundation fieldbus (FF) 基础现场总线

four connections 四连接

four level low current system worker 四级弱电工

FP (flexible point) (业务)灵活点

FPA (fire public address) 火灾事故广播

FPD (fire public device) 消防公共设备

FPGA (field-programmable gate array) 现场可编程门阵列

FPH (floor panel heating) 地板辐射采暖

FPH (Freephone) 话费总付电话，免费电话，被叫集中付费

fps (frames per second) 帧率单位(每秒帧数)

FR (frequency response) 频率响应

FR/LSOH (flame retardancy/low smoke zero halogen) 阻燃低烟无卤

fragmentation 分片

frame 帧，框架

frame check sequence (FCS) 帧校验序列

frame grabber 帧捕获器

frame rate 帧率

frame refresh frequency (FRF) 帧刷新频率，换帧频率

frame type 机框类型

frame type fire alarm controller 框式火灾报警控制器

frame-level equipment 机架级设备

frames per second (fps) 每秒帧数

framework bracket 支[骨]架

framing data 成帧数据

FRD (fire resistant damper) 防火阀

FRD (fire resisting damper) 防火阀

free access 自由入口

free space 空旷空间

Freephone (FPH) 话费总付电话，免费电话，被叫集中付费

freeze frame 冻结帧，预置点视频冻结

frequency (F) 频率

frequency allocation 频率配置

frequency conversion equipment 变频设备

frequency converter 频率变换器，变频器

frequency division multiple access (FDMA) 分频多址

frequency doubler 倍频器

frequency fluctuation rate 频率波动率

frequency mixer 混频器

frequency response (FR) 频率响应

frequency variation 频率变化

frequency-divided phase locking technology 分频锁相技术

fresh air supply (FAS) 新风供给

fresh air unit (FAU) 新风机组

F

FRF (frame refresh frequency) 帧刷新频率,换帧频率

front end system (FES) 前端系统

front room 前室

front screen projection 正[前]投影

front surface 正面

front upright post 前立柱

front-projected holographic display 全息投影

frost point 霜点

frozen earth depth 冻土深度

FRP (cable conduit of fiberglass reinforced plastic) 玻璃纤维增强塑料电缆导管

FRS (fire resistant shutter) 防火卷帘

FRT (fire retardant treatment) 阻燃处理

FSA (fiber to the service area) 光纤到服务区

FSK modulation FSK 调制

FSR (flame spread rate) 火焰传播速率

FSS (fixed satellite services) 固定卫星业务

FST (flame spread time) 火焰传播时间

FTP (file transfer protocol) 文件传输协议

FTP (foil twisted pair) 金属箔屏蔽双绞线电缆

FTTB (fiber to the building) 光纤到楼

FTTC (fiber to the curb) 光纤到路边

FTTD (fiber to the desk) 光纤到桌面

FTTF (fiber to the floor) 光纤到楼层

FTTH (fiber to the home) 光纤到户

FTTH (fiber to the house) 光纤到户

FTTN (fiber to the node) 光纤到节点

FTTO (fiber to the office) 光纤到办公室

FTTP (fiber to the premises) 光纤到驻地

FTTR (fiber to the remote) 光纤到远端

fuel cell 燃料电池

full contact connector (FC connector) FC 型光纤连接器

full duplex multipicture processing 全双工多画面处理

full shield 全屏蔽

full-scale amplitude 满度振幅

full-spectrum 全光谱

function 函数，功用，功能，函数过程，函（数）词

functional and protective earthing conductor 功能与保护接地导体

functional characteristic 功能特性

functional earth facility 功能接地装置

functional earthing 功能性接地

functional earthing conductor 功能接地导体

functional element 功能元素[元件]

functional entity (FE) 功能实体

functional integrity 功能完整性

functional performance 性能等级

fungal 真菌

fuse 保险管[丝]

fuse base 保险管座

fusion splice 熔接接头

fuzzy control 模糊控制

fuzzy logic 模糊逻辑

FWA (fixed wireless access) 固定无线接入

G

G frame G帧

G.711 G.711音频编码方式

G.722.1 G.722.1音频编码方式

G.726 G.726音频编码方式

gallon per minute (GPM) 加仑每分钟

galvanized sheet 镀锌板

galvanized steel tape 镀锌钢带

galvanize 镀锌

galvanized steel 镀锌钢(材)

gamma correction γ校正

ganged fading 联动衰减

garage 车库

gas fire detector 气体火灾探测器

gas fire extinguishing controller 气体灭火控制器

gas fire extinguishing device 气体灭火装置

gas shutoff valve 燃气关断阀

gas stove 燃气灶具

gate valve (GV) 闸阀

gateway mobile-services switching center (GMSC) 网关移动业务交换中心,关口MSC

gauge outfit 表头

GB picture GB图像

GC (generic cabling) 综合布线,通用布缆

GCS (generic cabling for customer premises) 用户建筑群的通用布缆

GCS (generic cabling system) 综合布线系统

GE16 (16 E1 interface board) GSM16路E1中继接口板

GECP (echo canceler pool) GSM回波抵消板,回波消除器池

general contractor 总承包

general inverter (GI) 通用变频器

general packet radio service (GPRS) 通用无线分组业务

generator room 发电机房

generic cabling (GC) 综合布线,通用布缆

generic cabling for customer premises (GCS) 用户建筑群的通用布缆

generic cabling system (GCS) 综合布线系统

generic cabling system for building and campus 建筑物与建筑群综合布线系统

generic telecommunications cabling 综合电信布线

genlock 同步锁相

geodetic chain 控制网

geographic information system (GIS) 地理信息系统

geometrical characteristic 几何特性

GG45 module GG45 模块

ghost (电视屏幕上的)重影

GI (general inverter) 通用变频器

GIF (graded-index fiber) 渐变光纤,渐变折射率光纤

GIMM (graded-index multimode) 渐变折射率多模光纤

girder 梁

GIS (geographic information system) 地理信息系统

glass beaded 玻珠幕

glass breakage detector 玻璃破碎探测器

glass fiber reinforced plastic rod 玻璃纤维增强塑料杆

glass fiber reinforced plastic duct 玻璃纤维增强塑料管

global system for mobile communications (GSM) 全球移动通信系统

glossy surface 光面

glowing combustion 灼热燃烧

glycol (or water) dry cooler 乙二醇(或水)干式冷却器

glycol (or water) free cooling fluid economizer cycle cooler 乙二醇(或水)自然循环节能冷却器

GMID (grandmaster identifier) 最高级时钟标识符

GMM/SM (GPRS mobility management and session management) GPRS 移动管理和会话管理

GMSC (gateway mobile-services switching centre) 网关移动业务交换中心,关口 MSC

GND (grounding) 接地

GNU general public license (GNU GPL) GNU 通用公共许可证

GNU GPL (GNU general public license) GNU 通用公共许可证

GOP (group of pictures) 画面组

GoS (grade-of-service) 服务级别

GPM (gallon per minute) 加仑每分钟

GPRS (general packet radio service) 通用无线分组业务

GPRS mobility management and

session management (GMM/SM) GPRS 移动管理和文字段落管理

GPRS radio resources service access point (GRR) GPRS 无线资源业务接入点

GPRS support node (GSN) GPRS 支持节点

G

GPRS tunneling protocol (GTP) GPRS 隧道协议

grade 等级评分

grade-of-service (GoS) 服务级别

graded-index fiber (GIF) 渐变光纤,渐变折射率光纤

graded-index optical fiber 渐变折射效率光纤

graded-index multimode (GIMM) 渐变折射率多模光纤

grandmaster identifier (GMID) 最高级时钟标识符

graph indicator in fire control center 消防控制中心图形显示装置

graphic management interface in Chinese 中文图形管理界面

graphical element 图形元素

graphical interface 图形接口

graphical representation 图示,图形表示法

graphics co-processor/accelerator 图形协处理器与加速键

graphics editor 图形编辑程序

graphics program 图形程序

grating optical fiber temperature fire detector 光栅光纤感温火灾探测器

gray scale 灰度

green building 绿色建筑

green power 绿色能源

grey scale reproduction 灰度等级重现

grille ceiling 格栅吊顶

GRM (circuit group monitor message) 电路群监视消息

gross profit (GP) 毛利

gross weight 毛重

ground clip 接地线夹

ground loop 接地环路

ground plate 接地板

ground source heat pump (GSHP) 地源热泵

ground wire 接地线

grounding (GND) 接地

grounding bar 接地排

grounding body 接地体

grounding connector 接地连接器

grounding copper bar 接地铜排

grounding fault 接地故障

grounding grid 接地网

grounding lead 接地引线

grounding plate combination 接地盘组合

grounding resistance 接地电阻

grounding system 接地系统

grounding terminal 接地端

group delay 群时延

group of pictures (GOP) 画面组

GRR (GPRS radio resources service access point) GPRS 无线资源业务接入点

GSHP (ground source heat pump) 地源热泵

GSM (global system for mobile communications) 全球移动通信系统

GSM mobile station (GSM MS) GSM 移动台

GSM MS (GSM mobile station) GSM 移动台

GSM Q3 protocol (Q3) GSM 的 Q3 协议

GSN (GPRS support node) GPRS 支持节点

GTP (GPRS tunneling protocol) GPRS 隧道协议

guard tour system 电子巡查系统

guide rail for erection 安装导轨

GV (gate valve) 闸阀

H

H.263　H.263 协议

H.264　H. 264 视频编解码技术标准

hacksaw　钢锯

hair hygrometer　毛发湿度计

half gain angle (HGA)　半增益视角

half rate (HR)　半速

half rate traffic channel (TCH/H)　半速率业务信道

half-duplex　半双工

half-duplex transmission (HDT)　半双工传输

hall　大厅

halogen-free low smoke and flame-retardant compound　无卤低烟阻燃电缆料

halogen-free flame-retardant optical fibre cable　无卤阻燃光缆

hand hole　手孔

handover　越区切换

hang up　挂机(电话)

hard ground　硬接地

harmful material　有害材料

harmonic　谐波

harmonic component　谐波分量

harmonic components in the current and voltage circuits　电流和电压电路中的谐波分量

harmonic content　谐波含量

HART (highway addressable remote transducer protocol)　可寻址远程传感器高速通道通信协议

hazardous energy level　危险能级

hazardous substances　有害物质

HBS (home bus system)　家庭总线系统

HC (home controller)　家庭控制器

HC (horizontal cross-connect)　水平交叉连接

HCA (hybrid channel assignment)　混合信道布置

HCS (hierarchical cell structure)　分层小区结构

HCS (higher order connection supervision)　高阶连接监督

HD (high definition) HD 图像格式

HD (horizontal distributor) 水平配线设备

HDA (horizontal distribution area) 水平配线区

HDB (HLR database) HLR 数据库

HDB3 (high density bipolar of order 3) 三阶高密度双极性码

HDCP (high-bandwidth digital content protection) 高带宽数字内容保护技术

HDCVI (high definition composite video interface) 高清复合视频接口

HDLC (high level data link control procedure) 高级数据链路控制(规程)

HDMI (high definition multimedia interface) 高清晰度数字多媒体(标准)接口

HDPE (high density polyethylene) 高密度聚乙烯

HDSL (high bit rate digital subscriber line) 高速率数字用户专用线路

HDT (half-duplex transmission) 半双工传输

HDTV (high definition television) 高清晰度电视

HDTVI (high definition transport video interface) (基于同轴电缆的)高清视频传输规范

HE (headend element) 前端控制器

headend 前[头]端,机头

headend element (HE) 前端控制器

heartbeat message 心跳消息

heat abstraction hole 散热孔

heat exchange system 热交换系统

heat exchanger unit for air conditioning and heating 空调和采暖换热机组

heat fire detector (HFD) 感温火灾探测器

heat island intensity 热岛强度

heat pump 热泵

heat release 散热

heat resistant ethylene-vinyl acetate rubber insulated cable 耐热乙烯-乙酸乙烯酯橡皮绝缘电缆

heat resistant silicone insulated cable 耐热硅橡胶绝缘电缆

heat shrinkable joint closure 热缩套管

heat source 热源

heating and cooling 加热和冷却

heating appliance 采暖设备

heating load 热负荷

heating panel 加[散]热板,辐射板
heating part 发热部件
heating system 供热系统
heating ventilating 供热通风
heating ventilation air conditioning (HVAC) 加热通风空调
helical scan 螺旋状扫描
help desk 帮助台
HEMS (home energy management system) 家居能源管理系统
HEPA (high efficiency particulate air filter) 高效过滤网络
HES (home electronic system) 家用电子系统
HEVC (high efficiency video coding) 高效率视频编码
hex nut 六角螺母
HF (high frequency) 高频
HFC (hybrid fiber coax) 混合光纤同轴电缆
HFC (hybrid fiber coaxial) 光纤同轴电缆混合网
HFD (heat fire detector) 感温火灾探测器
HFT (high-frequency transformer) 高频变压器
HG (home gateway) 家庭网关
HGA (half gain angle) 半增益视角
HGA (high gain antenna) 高增益天线
HIC (hybrid integrated circuit) 混合集成电路
Hi-color 高色彩
hierarchical cell structure (HCS) 分层小区结构
hierarchical control 分级控制
hierarchical network 分级[层]网,层次网
hierarchical routing 分级路由(选择)
hierarchical star topology 分层星型拓扑结构
hierarchical structure 递阶[层次]结构
HIFI (high fidelity) 高保真度音响
high birefringence optical fiber 高双折射光纤
high bit rate digital subscriber line (HDSL) 高速率数字用户专用线路
high definition (HD) HD 图像格式
high definition barcode image 高清码图
high definition composite video interface (HDCVI) 高清复合视频接口
high definition multimedia interface (HDMI) 高清数字多媒体(标

准)接口

high definition television (HDTV) 高清电视

high definition transport video interface (HDTVI) (基于同轴电缆的)高清视频传输规范

high density 高密度

high density bipolar of order 3 (HDB3) 三阶高密度双极性码

high density polyethylene (HDPE) 高密度聚乙烯

high efficiency particulate air filter (HEPA) 高效过滤网络

high efficiency video coding (HEVC) 高效率视频编码

high end 高端

high fidelity (HIFI) 高保真度音响

high frequency (HF) 高频

high gain antenna (HGA) 高增益天线

high gain power amplifier board (HPA) 高增益功放板

high impedance failure 高阻抗故障

high layer compatibility (HLC) 高层兼容性

high level data link control (procedure) (HDLC) 高级数据链路控制(规程)

high light compensation (HLC) 高亮度补偿,强光抑制

high loss fiber 高损耗光纤

high loss optical fiber 高损耗光纤

high memory area (HMA) 高端存储区

high potential difference (HPD) 高电位差

high power amplifier (HPA) 高[大]功率放大器

high sensitive detector 高灵敏型探测器

high speed backbone network (HSBN) 高速骨干网(络)

high speed local network (HSLN) 高速局域网(络)

high temperature 高温

high voltage (HV) 高电压

high working frequency 高工作频率

high-bandwidth digital content protection (HDCP) 高带宽数字内容保护技术

high-capacity mobile telecommunications system 大容量移动电话电信系统

higher order connection supervision (HCS) 高阶连接监督

higher order cross connect 高阶交叉连接

H

higher order path (HP) 高阶通路[通道]

higher order path adaptation (HOPA) 高阶通道适配

higher order path connection (HPC) 高阶通路连接

higher order path overhead monitor (HPOM) 高阶通路开销监视

high-frequency circuit board 高频电路板

high-frequency transformer (HFT) 高频变压器

high-humidity room 高湿度房间

high-rise dwelling building 高层住宅

high-speed data upload 高速数据上传

high-tech 高(新)技术

Highway Addressable Remote Transducer Protocol (HART) 可寻址远程传感器高速通道的开放通信协议

highway tunnel 公路隧道

hired and altered communication room 租房改建通信机房

history alarms 历史告警

history log 历史日志

HLC (high layer compatibility) 高层兼容性

HLC (high light compensation) 高亮度补偿,强光抑制

HLR (home location register) 归属位置注册处[寄存器]

HLR database (HDB) HLR 数据库

HLS (HTTP live streaming) HTTP 实时流媒体,(Apple)动态码率自适应技术

HMA (high memory area) 高端存储区

HMI (human machine interaction) 人机界面

HN (home network) 家庭网络

HOE (holographic optical element) 全息光学元件

hole site for erection 安装孔位

hollow area 镂空面积

holographic optical element (HOE) 全息光学元件

holographic screen 全息幕

home alarm system 家庭报警系统

home automation 家庭自动化

home bus system (HBS) 家庭总线系统

home controller (HC) 家庭控制器

home electronic system (HES) 家用电子系统

home energy management system (HEMS) 家居能源管理系统

home entrance 住宅入口

home exchange call 本局呼叫
home fire alarm controller 家用火灾报警控制器
home fire detector 家用火灾探测器
home gateway (HG) 家庭网关
home group 本群
home location register (HLR) 归属位置注册处[寄存器]
home network (HN) 家庭[本地]网络
home security 家居安防
home security system 家庭安全系统
home station 本站
home theater 家庭影院
honeycomb 蜂窝状(通风板)
hook function 钩子函数
hookup wire 布线用电线
hoop iron 大小喉箍
hop 跳跃,跳数
hop count 节点数,跳数
hop count limit 跳数限制值
hop limit 跳数限制
HOPA (higher order path adaptation) 高阶通道适配
hop-by-hop route 逐跳路由
horizontal and vertical sync pulse 行和场同步脉冲
horizontal built-in pipe 水平暗配管
horizontal cable 水平缆线[线缆]
horizontal cabling 水平布线
horizontal cabling subsystem 水平布缆子系统
horizontal cross-connect (HC) 水平交叉连接
horizontal distance 水平距离
horizontal distribution area (HDA) 水平配线区
horizontal distributor (HD) 水平配线设备
horizontal floor wiring 水平层布线电缆
horizontal installation 水平安装
horizontal scanning frequency (HSF) 行频,水平扫描频率
horizontal tilt 行[水平]倾斜
horizontal viewing angle 水平视角
horse power (HP) 马力
horsepower (HP) 马力
host 主机
hot dip 热浸镀锌
hot water supply system (HWSS) 热水供应系统
hotel 旅馆,宾馆
hot-line work 带电操作
hot-pluggable 热插拔,可带电插拔的
house mezzanine 夹层

household telephone call system 家庭电话呼叫系统

housing 护罩,外壳

howler tone 吼声,嗥鸣音

HP (higher order path) 高阶通路[通道]

HP (horse power/horsepower) 马力

HPA (high gain power amplifier board) 高增益功放板

HPA (high power amplifier) 高功率放大器

HPC (higher order path connection) 高阶通路连接

HPD (high potential difference) 高电位差

HPOM (higher order path overhead monitor) 高阶通路开销监视

HR (half rate) 半速

HRC (hypothetical reference circuit) 假设参考电路

HRDS (hypothetical reference digital section) 假设参考数字段

HRP (hypothetical reference path) 假设参考通道

HSBN (high speed backbone network) 高速骨干网

HSF (horizontal scanning frequency) 行频,水平扫描频率

HSLN (high speed local network) 高速局域网(络)

HTML (hypertext markup language) 超文本标记语言

HTTP (hypertext transfer protocol) 超文本传输协议

HTTP live streaming (HLS) HTTP 实时流媒体,(Apple)动态码率自适应技术

HTTPS (hypertext transfer protocol over secure socket layer) 安全套接字层超文本传输协议

HUB 集线器

human machine interaction (HMI) 人机界面

humidity 湿度

humidity ratio 含湿量

humidity sensor 湿度传感器

HV (high voltage) 高电压

HVAC (heating ventilation air conditioning) 加热通风空调

HWSS (hot water supply system) 热水供应系统

hybrid cable 混合线缆

hybrid channel assignment (HCA) 混合信道指配

hybrid circuit (2-wire/4-wire conversion) 二四线转换电路

hybrid circuit/network 混合电路网络

hybrid fiber coax (HFC) 光纤同轴

电缆混合网

hybrid fiber coaxial（HFC） 混合光纤同轴电缆

hybrid fiber coax access network 混合光纤同轴电缆接入网

hybrid integrated circuit（HIC） 混合集成电路

hybrid optical fiber cable 混合光缆

hybrid powered embedded thermal control equipment 交直流混合供电嵌入式温控设备

hybrid topology 混合拓扑

hybrid UPS（power）switch 混合UPS(电力)开关

hydraulic calculation 水力计算

hydrogen aging 氢老化

hypertext transfer protocol over secure socket layer（HTTPS） 安全套接字层超文本传输协议

hyperframe 超帧

hypertext markup language（HTML） 超文本标记语言

hypertext transfer protocol（HTTP） 超文本传输协议

hypothetical reference circuit（HRC） 假设参考电路

hypothetical reference digital section（HRDS） 假设参考数字段

hypothetical reference path（HRP） 假设参考通道

I

I picture　I 图像

I/O (input/output)　输入输出端口

IAS (intruder alarm system)　入侵报警系统

IB (intelligent building)　智能建筑[楼宇]

IBCS (intelligent building cabling system)　智能建筑[楼宇]布线系统

IBDN (integrated building distribution network)　楼宇综合布线网络(美国百通公司综合布线品牌名)

IBMS (intelligent building management system)　智能建筑[楼宇]管理系统

IBN (isolated bonding network)　独立的联结网络

IBR (image-based rendering)　基于图形的绘制

IBS (intelligent building system)　智能建筑[楼宇]系统

IC (integrated circuit)　集成电路

IC (interaction channel)　交互通道

IC (interlock code)　闭锁码

IC (intermediate cross-connect)　中间交叉连接

IC extractor　IC 起拔器

ICMP (Internet control message protocol)　因特网控制信息协议

ICP (Internet content provider)　因特网内容提供商

ICR (IR-cut removable)　双滤光片切换器,日夜转换

ICS (interference cancellation system)　干扰消除系统

ICT (information and communications technology)　信息和通信技术

ID (intermediate distributor)　中间配线设备,中间配线架

IDC (insulation displacement connector)　绝缘位移连接器

IDC (Internet data center)　因特网数据中心

IDE (integrated drive electronics)　电子集成驱动器

identification mark　识别标志

identifier 标识符

IDF (intermediate distribution frame) 楼层配线架,分配线架

idle channel 空闲信道

IDS (industry distribution system) 工业布线系统

IDS (intrusion detection system) 入侵检测系统

IEC (International Electrotechnical Commission) 国际电工委员会

IEC jack IEC 插孔

IEC plug IEC 插头

IEEE (Institute of Electrical and Electronics Engineers) 电气及电子工程师学会

IEH (indirect electric heating) 间接电加热

IETF (Internet Engineering Task Force) 因特网工程任务组

IF (intermediate frequency) 中频

IGMP (Internet group management protocol) 网际组管理协议,因特网组管理协议

ignition source 引火源

ignition temperature 引燃温度

IGS (inert gas system) 惰性气体系统

IID (industrial intermediate distributor) 工业中间配线架

IIS (intelligent integration system) 智能化集成系统

IIS (Internet information service) 因特网信息服务

IL (insertion loss) 插入损耗

ILD (insertion loss deviation) 插入损耗偏差

illumination 光照度

ILS (intelligent lighting system) 智能灯光[照明]系统

image 概念,图像,镜像,映像

image based rendering (IBR) 基于图形的绘制

image display 图像显示

image retention 图像残留

image signal processing (ISP) 图像信号处理器

image transmission 图像传输

image type flame detector 图像型火焰探测器

IMAP (Internet message access protocol) 因特网信息访问协议

IMEI (international mobile equipment identity) 国际移动设备身份码

IMEI (international mobile station equipment identity) 国际移动台设备标识,国际移动设备标识

IMGI (international mobile group identity) 国际移动组标识

immediate hotline 立即[即时]热线

impact 冲击力
impact resistance 耐冲击性
impact sound pressure level 撞击声压级
impedance 阻抗
impedance matching adapter 阻抗匹配适配器
implementation under test (IUT) 被测系统
import and export control (access control) system 出入口控制(门禁)系统
import Customs clearance 进口通关
important user data 重要用户数据
IMSI (international mobile subscriber identity) 国际移动用户标识
IMU (intelligent management unit) 智能管理单元
in service software upgrade (ISSU) 服务软件升级
IN switching management (IN-SM) IN 交换管理
inactive link 非活动链
INAP (intelligent network application protocol) 智能网应用规程
in-band spectrum ripple 带内频谱不平坦度
incandescent lamp 白炽灯
INCM (intelligent network conceptual model) 智能网概念模型
incoming 来话
incoming call 来话[入局]呼叫
INCS-1 (intelligent network capability set-1) 智能网能力集第一阶段
independent 独立
independent third-party test institution 独立第三方测试机构
in-depth design 深化设计
index of thermal inertia 热惰[惯]性指标
indicator 指示器
indicator light 指示灯
indirect 间接
indirect DC convertor 间接直流变流器
indirect electric heating (IEH) 间接电加热
indirect grounding 间接接地
individual harmonic distortion 单次谐波畸变
individual layer 个体层
individual work area 单独[独立]工作区
individually screened pair 线对[对对]屏蔽
indoor bushing 室内[户内]套管
indoor cable 室内[户内]缆线
indoor covering system 室内[户

内]覆盖系统

indoor external insulation 室内[户内]外绝缘

indoor multi-mode 10G optical fiber cable 室内[户内]多模万兆光缆

indoor optical fibre 室内[户内]光缆

indoor optical fibre cable 室内[户内]光缆

indoor positioning system (IPS) 室内[户内]定位系统

indoor signal distributing system (无线通信)室内[户内]信号分布系统

indoor single-mode optical fiber cable 室内[户内]单模光缆

indoor telecom distribution pipes network 室内[户内]电信配线管网

indoor temperature 室内[户内]温度

indoor type 室内[户内]型

indoor unit 室内[户内]机

indoor-immersed bushing 室内[户内]浸入式套管

inductance 电感

induction card 感应卡

inductive charge 感应电荷

inductor 电感器

industrial communications network 工业通信网络

industrial computer (IPC) 工业控制微机,工控机

industrial computer and the chassis 工控机及其机箱

industrial environment cabling system 工业环境布线系统

industrial frequency inductor 工频电感器

industrial horizontal ruler 工业水平尺

industrial intermediate distributor (IID) 工业中间配线架

industrial machinery 工业机械

industrial personal computer (IPC) 工业个人计算机

industrial premises 工业建筑群

industrial standard architecture (ISA) bus 工业标准结构总线

industrial video device 工业视频装置

industry distribution system (IDS) 工业布线系统

inert gas fire-fighting 惰性气体灭火

inert gas fire-fighting system 惰性气体灭火系统

inert gas system (IGS) 惰性气体系统

inerting concentration 惰化浓度

inerting system 惰化系统

information and communications technology (ICT) 信息和通信技术

information appliance by embedded processors 含嵌入式处理器的信息家电

information application system 信息化应用系统

information distribution box 信息配线箱

information facility system 信息设施系统

information highway 信息高速公路

information module 信息模块

information network system (INS) 信息网络系统

information outlet (IO) 信息插座

information security technology (IST) 信息安全技术

information service 信息服务

information technology (IT) 信息技术

information technology cabling 信息技术布缆

information technology equipment (ITE) 信息技术设备

information transfer 信息传输

infrared beam 红外光束

infrared light sensor smoke detector 红外光束感烟火灾探测器

infrared port 红外端口

infrared sensor 红外感应器

infrared temperature measurement of electrical fire monitoring detector 红外测温式电气火灾监控探测器

infrastructure 基础设施

initial channel assignment 初始化信道分配

initial design 初始设计

initial effective lumen 初始有效光通量

initial inspection 初验

initial luminaire efficacy 初始光效

injection 喷射

inner conductor 内导体

inner electrostatic potential 内静电势,室内静电电位

inorganic matter 无机物

input current distortion 输入电流畸变

input frequency tolerance 输入频率允差

input power factor 输入功率因数

input voltage distortion 输入电压畸变

input voltage susceptibility of AES/EBU interface AES/EBU 接口

输入电压灵敏度
input voltage tolerance 输入电压允差[容差]
input/output (I/O) 输入输出(端口)
input/output terminal (IOT) 输入输出终端
INS (information network system) 信息网络系统
insertion gain 插入增益
insertion loss (IL) 插入损耗
insertion loss deviation (ILD) 插入损耗偏差
insertion signal 插入信号
insertion test signal (ITS) 插入测试信号
IN-SM (IN switching management) IN 交换管理
inspection 检验
inspection record 检验记录
inspection report 检验报告
inspection report of complete unit 整机检测报告
inspector 检验方
installation 安装
installation accessory 安装配件
installation and construction unit 安装施工单位
installation coupler 安装式耦合器
installation detail drawing 安装详图
installation method 安装方法
installation parts 安装件
installation planning 安装规划
installation practice 安装实践
installation process 安装流程
installation spacing 安装间距
installation specification 安装规范
installation subcontracting institution (设备)安装分包机构
installed capacity 装机容量
installer 安装者
installing support 安装支架
instant alarm 即时告警
Institute of Electrical and Electronics Engineers (IEEE) 电气及电子工程师学会
instruction 指示,指令
instruction manual 产品说明书
insulated conductor 绝缘导体
insulated wire 绝缘导线
insulating material 绝缘材料
insulation 绝缘
insulation covering 绝缘套
insulation displacement connection (IDC) 绝缘位移连接
insulation displacement connector (IDC) 绝缘位移连接器
insulation impedance 绝缘阻抗
insulation layer 绝缘层

insulation pad 绝缘垫
insulation piercing connection (IPC) 绝缘刺穿连接
insulation resistance 绝缘电阻
INT (integration) 一体化
integrated building distribution network (IBDN) 楼宇综合布线网络(美国百通公司综合布线品牌名)

integrated circuit (IC) 集成电路
integrated drive electronics (IDE) 电子集成驱动器
integrated lightning protection system 综合防雷保护系统
integrated part load value (IPLV) 综合部分负荷性能系数
integrated resistor 集成电阻器
integrated services digital network (ISDN) 综合业务数字网(络)
integrated services local area network (ISLAN) 综合业务局域网(络)
integrated software 集成软件
integrated surveillance 集成监控
integrated surveillance center (ISC) 集成监控中心
integration (INT) 一体化
integrity 完整性
intellectual property right (IPR) 知识产权
intelligence 智能特性
intelligent 10-pin patch cord 10针智能跳线
intelligent building (IB) 智能建筑[楼宇]
intelligent building cabling system (IBCS) 智能建筑[楼宇]布线系统
intelligent building management system (IBMS) 智能建筑[楼宇]管理系统
intelligent building system (IBS) 智能建筑[楼宇]系统
intelligent community system 智能社区系统
intelligent control system structure 智能控制系统结构
intelligent FO patch cord 光纤智能跳线
intelligent front end system 智能前端系统
intelligent household appliance 智能家用电器
intelligent integration system (IIS) 智能化集成系统
intelligent lighting system (ILS) 智能灯光[照明]系统
intelligent management 智能管理
intelligent management unit (IMU) 智能管理单元
intelligent monitoring area 智能监

控领域

intelligent monitoring center 智能化管理中心

intelligent network application protocol (INAP) 智能网应用规程

intelligent network capability set-1 (INCS-1) 智能网能力集第一阶段

intelligent network conceptual model (INCM) 智能网概念模型

intelligent patch cord 智能跳线

intelligent patch panel (IPP) 智能配线架

intelligent platform management interface (IPMI) 智能平台管理接口

intelligent service system (INTESS) 智能业务系统

intelligent sunshading system/electric curtain 智能遮阳系统(电动窗帘)

intelligent system in household 家居智能化系统

intelligent system in residential district 住宅小区智能化系统

intelligent system integrated (ISI) 智能化系统集成

intelligent terminal 智能终端

intelligent transmission 智慧传输

intelligent transmission cloud 智慧传输云

intelligentization technology 智能化技术

intended life 预期寿命

inter coding 帧间编码

inter conference 交互式会议

inter prediction 帧间预测

interaction channel (IC) 交互通道

interactive broadcast 交互广播

interactive cable TV 交互式有线电视

interactive network (Internet) 交互式网络(因特网)

interactive processing 交互式处理

interactive program guide (IPG) 交互节目指南

interactive television (ITV) 交互电视

interactive video telegraphy videotex 交互式可视图文

interactive video-on-demand (IVOD) 交互式视频点播

interactive voice response (IVR) 交互式语音应答

intercell handover 小区间切换

intercom system 对讲系统

interconnect 互连,互联

interconnectability 可互联性

interconnected independent fire detector 互联型独立式火灾探

测器

interconnection 对接,互连,互联

interconnection and mutual control 互联互控

inter-connection bolt 互连螺栓

interconnection subsystem (IS) 互联系统

interface 接口,界面

interface card 接口板[卡]

I

interface circuit chip 接口电路芯片

interface ID 接口标识符

interface power driver 接口功率驱动器

interference cancellation system (ICS) 干扰消除系统

interference-adaptive system 干扰自适应系统

interlace 隔行

interlace ratio 隔行比

interlaced scanning 隔行扫描

interleave 隔行扫描

interlock 内锁

interlock code (IC) 闭锁码

intermediate cable 中间线缆

intermediate cabling subsystem 中间布缆子系统

intermediate cross-connect (IC) 中间交叉连接

intermediate distribution frame (IDF) 楼层配线架,分配线架

intermediate distributor (ID) 中间配线设备,中间配线架

intermediate frequency (IF) 中频

intermediate frequency amplifier 中频放大器

intermediate system to intermediate system (IS-IS) 中间系统到中间系统

intermittent fault 间歇故障

internal cable 内部线缆

internal cycle 内循环

internal user data 内部用户数据

Internation Special Committee on Radio Interference (CISPR) 国际无线电干扰特别委员会

international annealed copper standard (IACS) 国际退火铜标准

International Electrotechnical Commission (IEC) 国际电工技术委员会

International Gateway 国际通信进出口局

international mobile equipment identity (IMEI) 国际移动设备身份码

international mobile group identity (IMGI) 国际移动组标识

international mobile station equipment identity (IMEI) 国际移动台设备标识,国际移动设备标识

international mobile subscriber identity (IMSI) 国际移动用户标识

International Organization for Standardization (ISO) 国际标准化组织

International Radio Consultative Committee (CCIR) 国际无线电咨询委员会

international standard 国际标准

International Telecommunication Union (ITU) 国际电信联盟

International Telecommunication Union-Telecommunication Standard Sector (ITU-T) 国际电信联盟-电信标准部

Internet 互联网，因特网

Internet company 因特网企业

Internet content provider (ICP) 因特网内容提供商

Internet control message protocol (ICMP) 因特网控制信息协议

Internet data center (IDC) 因特网数据中心

Internet Engineering Task Force (IETF) 因特网工程任务组

Internet group management protocol (IGMP) 网际组管理协议，因特网组管理协议

Internet information service (IIS) 因特网信息服务

Internet message access protocol (IMAP) 因特网信息访问协议

Internet of Things (IoT) 物联网

Internet Protocol (IP) IP协议，因特网协议

Internet Protocol multicast (IP-M) IP广播业务

Internet Protocol television (IPTV) 交互式网络电视

Internet Protocol version 4 (IPv4) 因特网协议(Internet Protocol，IP)的第四版

Internet Protocol version 6 (IPv6) 因特网协议(Internet Protocol，IP)的第六版

Internet Society (ISOC) 国际互联网学会

Internet Streaming Media Alliance (ISMA) 国际互联网流媒体联盟

interoperability 可互操作性

interpolated prediction 插值预测

interpolation 插值法

inter-process communication (IPC) 进程间通信

interrupt disable 禁止中断

interruption time 中断时间

inter-symbol interference (ISI) 符号间干扰，码间干扰

I

interval 间隔
interworking facilities 互通设施
interworking function (IWF) 互通功能
interworking system 互通系统
interworking unit 互通部件[单元]
INTESS (intelligent service system) 智能业务系统

intimidation alarm system 胁迫报警系统
intra-prediction 帧内预测
intruder alarm system (IAS) 入侵报警系统
intrusion detection system (IDS) 入侵检测系统
intrusion prevention system (IPS) 入侵预防系统
invalid cell 无效信元
in-vehicle network 车载网络
inverse phase 反相位
inverse transform 反变换
inverter 变频器
IO (information outlet) 信息插座
IOT (input/output terminal) 输入输出终端
IoT (Internet of Things) 物联网
IP (internet protocol) 因特网协议,因特网协议
IP bypass IP 旁路
IP cable modem IP 电缆调制解调器
IP camera (IPC) IP 摄像机
IP multicasting technology IP 多路广播技术
IP phone IP 电话
IP router IP 路由器
IP storage area network (IP-SAN) 基于 IP 的存储局域网络
I-passive optical network (I-PON) I 无源光网络
IPC (industrial computer) 工业控制微机,工控机
IPC (industrial personal computer) 工业个人计算机
IPC (insulation piercing connection) 绝缘刺穿连接
IPC (inter-process communication) 进程间通信
IPC (IP camera) IP 摄像机
IPG (interactive program guide) 交互节目指南
IPLV (integrated part load value) 综合部分负荷性能系数
IP-M (internet protocol multicast) IP 广播业务
IPMI (intelligent platform management interface) 智能平台管理接口
I-PON (I-passive optical network) I 无源光网络

IPP (intelligent patch panel) 智能配线架

IPR (intellectual property right) 知识产权

IPS (indoor positioning system) 室内定位系统

IPS (intrusion prevention system) 入侵预防系统

IP-SAN (IP storage area network) 基于IP的存储局域网络

IPTV (Internet Protocol television) 交互式网络电视

IPv4 (Internet Protocol version 4) 因特网协议(Internet Protocol, IP)的第四版

IPv6 (Internet Protocol version 6) 因特网协议(Internet Protocol, IP)的第六版

IR-cut removable (ICR) 双滤光片切换器,日夜转换

irradiation 照射

IS (interconnection subsystem) 互联系统

ISC (integrated surveillance center) 集成监控中心

ISDN (integrated services digital network) 综合业务数字网

ISDN primary rate interface (ISDN-PRI) ISDN基群速率接口

ISDN station set ISDN话机

ISDN user part (ISUP) 综合业务数字网用户部分

ISDN user part (SS7) (ISUP) ISDN用户部分(七号信令)

ISDN-PRI (ISDN primary rate interface) ISDN的基群速率接口

ISI (intelligent system integrated) 智能化系统集成

ISI (inter-symbol interference) 符号间干扰,码间干扰

IS-IS (intermediate system to intermediate system) 中间系统到中间系统

ISLAN (integrated services local area network) 综合业务局域网

ISMA (Internet Streaming Media Alliance) 国际互联网流媒体联盟

ISO (International Organization for Standardization) 国际标准化组织

ISOC (Internet Society) 国际互联网学会

isolated bonding network (IBN) 独立的联结网络

isolated lightning protection system 孤立防雷保护系统

isolation 隔离度

isolator 隔离器

isometric drawing 轴测图,等角

图，正等轴测图

ISP (image signal processing) 图像信号处理器

ISSU (in service software upgrade) 服务软件升级

IST (information security technology) 信息安全技术

ISUP (ISDN user part SS7) ISDN 用户部分(七号信令)

ISUP (ISDN user part) 综合业务数字网用户部分

IT (information technology) 信息技术

IT equipment IT 设备

IT product 信息技术产品

ITE (information technology equipment) 信息技术设备

item 条目

item-by-item substantive response 逐项实质性应答

itemized inspection report 分项检测报告

ITS (insertion test signal) 插入测试信号

ITU (International Telecommunication Union) 国际电信联盟

ITU-T (International Telecommu-nication Union-Telecommunica-tion Standard Sector) 国际电信联盟[国际电联]电信标准部

ITV (interactive television) 交互电视

IUT (implementation under test) 被测系统

IVOD (interactive video-on-demand) 交互式视频点播

IVR (interactive voice response) 交互式语音应答

IWF (interworking function) 互通功能

J

jacket 套壳,护套

JBOD (just a bunch of disks) 磁碟簇

jitter 抖动,不稳定性

jitter susceptibility 抖动灵敏度,抖动敏感性

JoinNet web meeting 多媒体视讯会议系统

Joint Technical Committee (JTC) 联合技术委员会

Joint Video Team (JVT) 联合视频组

joint-box 电缆接线箱

JTC (Joint Technical Committee) 联合技术委员会

judgment 判断

jump wire 飞线,跳线

jumper 跳线,压接跳线

jumper cable 跳线电缆

jumper wire 跳[跨接]线

jumper wire manager 跳线管理器

junction box 分线箱,接线盒

just a bunch of disks (JBOD) 磁碟簇

JVT (Joint Video Team) 联合视频组

K

keep away　避开

key　关键值;键;密钥;色控值;钥匙

key frequency　关键频率

key user data　关键用户数据

keyboard drawer　键盘抽屉

keyboard video mouse (KVM)　键盘视频鼠标,多计算机切换器

key-frame interval　关键帧间隔

keying　锁键

keystone correction　梯形校正,梯形失真[畸变]校正

keyword　关键词

kilowatt (kW)　千瓦

kitchen　厨房

Krone　综合布线产品品牌名

Kvar　千乏

KVM (keyboard video mouse)　键盘视频鼠标,多计算机切换器

kW (kilowatt)　千瓦

kWh　千瓦时

L

L2TP (layer 2 tunneling protocol) 第二层隧道协议

label 标签,标记

label box made of organic glass 有机玻璃标签框

label description 标识说明

label edge router (LER) 标记边缘路由器

label frame of polymethyl methacrylate 聚甲基丙烯酸甲酯标签架,有机玻璃标签框

label of subject and object 主客体标记

label switch router (LSR) 标记交换路由器

labelling strip 标签条

LAC (location area code) 位置区号码

LACP (link aggregation control protocol) 链路汇聚控制协议

laminated metal plastic foil 金属塑料复合箔

LAN (local area network) 局域网

LAN emulation server (LES) 局域网仿真服务器

LAN equipment (local area network equipment) 局域网设备

LAP (link access procedure) 链路访问过程

LAPB (link access procedure balanced for X.25) X.25 平衡链路访问规程

LAPD (link access procedure protocol for D channel) D 信道链路访问协议

largest coding block 最大编码块

largest coding unit (LCU) 最大编码单元

LASER (light amplification by stimulated emission of radiation) 受激辐射光放大,激光器

laser diode 激光二极管

laser source 激光光源

laser vision disc (LD) 激光影碟机,激光视盘

LAT (leaving air temperature) 出

风温度

LATA (local access and transport area) 本地访问和传输区域

latent fault 潜在故障

launcher 发射器

layer 2 tunneling protocol (L2TP) 第二层隧道协议

LBC (load bearing capacity) 承载能力

LC (line concentrator) 集线器

LC (lucent connector) LC 型光纤连接器

LC connector LC 光纤连接器

LC coupler LC 耦合器

LC fiber pigtail LC 光纤尾纤

LC interface LC 接口

LCD (liquid crystal display) 液晶显示屏

LCD connector 双芯 LC(光纤)连接器

LCL (longitudinal conversion loss) 纵向转换损耗

LCL (longitudinal to differential conversion loss) 纵向差分转换损耗

LC-LC multi-mode optical fiber jumper wire LC-LC多模光纤跳线

LCOS (liquid crystal on silicon) 液晶附硅

LCP (link control protocol) 链路控制协议

LCTL (longitudinal to differential conversion transfer loss) 纵向差分转换传送损耗

LCU (largest coding unit) 最大编码单元

LD (laser vision disc) 激光影碟机,激光视盘

LDAP (lightweight directory access protocol) 轻量目录访问协议

LDP (local distribution point) 局部配线点

LDPC (low density parity check code) 低密度奇偶校验码

LDPE (low density polyethylene) 低密度 PE,低密度聚乙烯

LE (local exchange) 本地交换网[局]

lead antenna 引向天线

leak out acoustic attenuation 漏出声衰减

leakage alarm 漏报警

leaky cable 漏泄电缆

leaky cable radio communication system 漏泄电缆无线通信系统

least significant bit (LSB) 最低有效位

leaving air temperature (LAT) 出风温度

leaving water temperature (LWT) 出水温度
LEC (local exchange carrier) 本地交换运营商,本地交换电信公司
LED (light emitting diode) 发光二极管
LED display screen 发光二极管显示屏
LED source LED光源
lens 镜头
lens shade 镜头遮光罩
LER (label edge router) 标记边缘路由器
LES (LAN emulation server) 局域网仿真服务器
level difference between left and right channels 左右声道电平差
level gauge 液位计
level of digitally modulated signal 数字调制信号电平
level of protection 防护等级
level of risk 风险等级
level of security 安全防护水平
level range of radio programme 广播节目的电平范围,电平均衡
level up degree 平整度
liaison meeting 联络会(议)
life cycling 生命周期
life expectancy 预期[使用]寿命
lift cable 电梯电缆
light amplification by stimulated emission of radiation (LASER) 受激辐射光放大,激光器
light communication 光通信
light emitting diode (LED) 发光二极管
light flux 光通量
light intensity 光强度
light leak 漏光
lighting 照明
lighting control 灯光控制
lighting controller 照明控制器
lighting device 照明装置
lighting protection zone (LPZ) 闪电[雷电]防护区
lightness 光亮度
lightning 闪电
lightning grounding clip 防雷接地夹
lightning protection 防雷,雷电防护
lightning protection system (LPS) 防雷保护系统
lightning strike 雷击
lightweight directory access protocol (LDAP) 轻量目录访问协议
limit curve 极限曲线
limit curve of critical state 临界极限曲线
limited current circuit 限流电路

limiting value 限值

line compensation 线路补偿

line concentrator (LC) 线路集中器

line guarding 线警戒

line location 线位

line number 线号

line of sight 视线

linear beam smoke fire detector 线型光束感烟火灾探测器

linear fire detector 线型火灾探测器

linear load 线性负载

linear pulse code modulation (LPCM) 线性脉冲编码调制

linearization distortion 线性化失真

line-type fire detector 线型火灾探测器

link 连接,链路,链接

link access procedure (LAP) 链路访问过程

link access procedure balanced for X.25 (LAPB) X.25平衡链路访问规程

link access procedure protocol for D channel (LAPD) D信道链路访问协议

link aggregation control protocol (LACP) 链路汇聚控制协议

link circuit 链路

link control protocol (LCP) 链路控制协议

link length 链路长度

link offset 链路偏移量

link state advertisement (LSA) 链路状态通告

link subsystem 干线子系统

linkage control design 联动控制设计

linkage control mode 联动控制方式

linkage function 联动功能

linkage operation 联动操作

lintel 过梁

lip sync 唇音同步技术

liquefied petroleum gas (LPG) 液化石油气

liquid crystal display (LCD) 液晶显示屏

liquid crystal on silicon (LCOS) 液晶附硅

LIS (logical IP subnet) 逻辑IP子网

list of response to the technical requirement 技术要求响应表

liter per minute (LPM) 每分钟公升(水流量)

literal mark strip 文字标识条

live broadcast system 现场直播

系统

live data collection 实时数据采集

live QR Code 二维码活码,实时二维码

LKFS (Loudness, K-weighted, relative to full scale) 响度的单位、K加权、相对于全尺度

LLASM (location lock application software module) 位置锁定应用软件模块

LLC (logic link control layer) 逻辑链路控制层

LLC (logic link control) 逻辑链路控制

LLMI (location lock module identification) 位置锁定模块识别号

LMF (lumen maintenance factor) 流明维持因子,光通维持率

LMI (local management interface) 本地管理接口

LNS (look n stop) LNS防火墙(法国)

load 装入,负载

load balance 负载均衡

load bearing capacity (LBC) 承载能力

load control / restriction 负荷控制限制

load power factor 负载功率因数

load resistance 负载阻抗

load voltage 负载电压

local access and transport area (LATA) 本地访问和传输区域

local area network (LAN) 局域网

local area network equipment (LAN equipment) 局域网设备

local distribution point (LDP) 局部配线点

local distribution space 本地配线空间

local exchange (LE) 本地交换网[局]

local exchange carrier (LEC) 本地交换运营商,本地交换电信公司

local management interface (LMI) 本地管理接口

local service provider (LSP) 本地服务提供商

local telecommunication cable 市内通信电缆

location area code (LAC) 位置区号码

location lock 位置锁定

location lock application software module (LLASM) 位置锁定应用软件模块

location lock mode 位置锁定模式

location lock module 位置锁定模块

location lock module identification (LLMI) 位置锁定模块识别号
location unlock mode 位置解锁模式
locking bar system 锁杆系统
locking mechanism 锁定机构
LOF (loss of frame) 帧丢失
logic 逻辑关系
logic link control (LLC) 逻辑链路控制
logic link control layer (LLC layer) 逻辑链路控制层
logical IP subnet (LIS) 逻辑 IP 子网
logo 标识
logo generator 标识发生器
LON LonWorks 协议
long-term evolution (LTE) 长期演进
longitudinal conversion loss (LCL) 纵向转换损耗
longitudinal conversion transfer loss 纵向转换传送损耗
longitudinal to differential conversion loss (LCL) 纵向差分转换损耗
longitudinal to differential conversion transfer loss (LCTL) 纵向差分转换传送损耗
longitudinal wrap 纵包
longitudinal-depth protection 纵深防护
longitudinal-depth protection system 纵深防护体系
LonWorks LonWorks 协议
lood vacuum pump 真空泵
look n stop (LNS) LNS 防火墙(法国)
loop 循环,回路,环
loop filter 环路滤波
loop play 循环播放
loop resistance 回路电阻
loose tube 松套管
loss of administrative unit pointer (AU-LOP) 管理单元指针丢失
loss of frame (LOF) 帧丢失
loss of synchronization of a picture 图像同步丢失
loss tolerance 损耗容限
lost packet recovery (LPR) 丢包恢复
loudness 音量,响度
loudness level 响度级
Loudness, K-weighted, relative to full scale (LKFS) 响度的单位
loudspeaker (LS) 扬声器,音箱
low birefringence optical fiber 低双折射光纤
low current system engineering 弱电工程
low current system worker 弱电工

low density parity check code (LDPC) 低密度奇偶校验码
low density polyethylene (LDPE) 低密度PE,低密度聚乙烯
low frequency signal 低频信号
low impedance failure 低阻抗故障
low impedance termination 低阻抗终接
low pressure switch 低压压力开关
low smoke and halogen-free (LSOH) 低烟无卤
low smoke and halogen-free material 低烟无卤材料
low temperature place 低温场所
low voltage distribution system (LVDS) 低压配电系统
low voltage system 弱电系统
lower order path (LP) 低阶通道
lower order path overhead monitor (LPOM) 低阶路径开销监视
low-rise dwelling building 低层住宅
low-smoke halogen-free (LSOH) 低烟无卤
low-voltage (LV) 低电压
LP (lower order path) 低阶通道
LPCM (linear pulse code modulation) 线性脉冲编码调制
LPG (liquefied petroleum gas) 液化石油气
LPM (liter per minute) 每分钟公升(水流量)
LPOM (lower order path overhead monitor) 低阶路径开销监视
LPR (lost packet recovery) 丢包恢复
LPS (lightning protection system) 防雷保护系统
LPZ (lightning protection zone) 闪电[雷电]防护区
LS (loudspeaker) 扬声器,音箱
LSA (link state advertisement) 链路状态通告
LSA plus terminal block LSA plus端接模块
LSB (least significant bit) 最低有效位
LSOH (low smoke and halogen-free) 低烟无卤
LSOH (low-smoke halogen-free) 低烟无卤
LSP (local service provider) 本地服务提供商
LSR (label switch router) 标记交换路由器
LTE (long-term evolution) 长期演进
lucent connector (LC) LC型光纤连接器
luma 亮度

lumen 流明

lumen maintenance factor (LMF) 流明维持因子,光通维持率

luminance 亮度

luminance brightness 光亮度

luminance crosstalk 亮度串扰

luminance noise 亮度噪声

luminous flux 光通量

LV (low-voltage) 低电压

LVDS (low voltage distribution system) 低压配电系统

LWT (leaving water temperature) 出水温度

LZH connector LZH(光纤)连接器

L

M

M12　M 12 接口

MA (management automation)　管理自动化

MAC (mandatory access control)　强制访问控制

MAC (media access control)　媒体[介质]访问控制(子层)

MAC domain　MAC 域

machine cycle (MC)　机器周期

machine language　机器语言

macro-bend loss (MBL)　宏弯损耗

macro-bending　宏弯曲

macro-bending characteristics　宏弯特性

macro-module　宏模块,宏块

MADI (multi-channel audio digital interface)　多通道音频数字串行接口

MAG (magnet)　磁铁

MATV (master antenna television)　主天线电视,共用天线电视

Mag Lock (magnetic lock)　电磁锁

magnet (MAG)　磁铁

magnetic contact　磁开关

magnetic lines of force　磁力线

magnetic lock (Mag Lock)　电磁锁

magnetic shielding　磁屏蔽

magnetic strip　磁条

main contractor　主承包方

main cross-connect (MC)　主交叉连接

main distributing facility (MDF)　主配线设施

main distribution area (MDA)　主配线区

main distribution frame (MDF)　主配线架

main distributor (MD)　主配线架

main earthing busbar　总接地母线

main earthing terminal (MET)　总接地端子

main logic board (MLB)　主逻辑板

main module application software　主模块应用软件

main power cabling　主干电力布缆

main power supply system　主供电

系统

maintenance bypass 维修旁路

maintenance hole 人孔

maintenance of polarity 极性维护

maintenance system 维护保养制度

major installation contractor 安装主承包

make-time 闭合时间

mall 商场

MAN (manual) 手册,说明书

MAN (metropolitan area network) 城域网

managed object (MO) 被管理对象

managed object relationship 被管理对象关系

management automation (MA) 管理自动化

management database (MDB) 管理数据库

management domain (MD) 管理领域

management information base (MIB) 管理信息库

management information system (MIS) 管理信息系统

management inhibit message (MIM) 管理阻断消息

management method 管理方法

management subsystem 管理子系统

management system 管理系统

mandatory access control (MAC) 强制访问控制

manual (MAN) 手册,说明书

manual control (MC) 手控,手动控制

manual control panel 手动控制盘

manual fire alarm button 手动火灾报警按钮

manual fire alarm call point 手动火灾报警呼叫点

manual start button 手动启动按钮

manual stop button 手动停止按钮

manual white balance (MWB) 手动白平衡

manufacturing process 生产流程

margin 余量

marking system 标识系统

master antenna television (MATV) 主天线电视,共用天线电视

master control system (MCS) 播控系统,主控系统

matched filter (MF) 匹配滤波器

mated pair 匹配线对

material 材料,素材

matt 亚光

maximum allowable level 最大允许电平

maximum beam range 最大射束距离

maximum ramp-down time 最大下降时间

maximum ramp-up time 最大上升时间

maximum transmission unit (MTU) 最大传输单元

MBGP (multiprotocol BGP) 组播协议[多协议]边界网关协议

MBL (macro-bend loss) 宏弯损耗

MC (machine cycle) 机器周期

MC (main cross-connect) 主交叉连接

MC (manual control) 手控，手动控制

MCA-STREAM (multi-channel audio stream) 多通道数字音频传输技术，多声道音频流

MCCS (multimedia collaborative conference system) 多媒体协同会议系统

MCL (minimum coupling loss) 最小耦合损耗

MCONF (multimedia conferencing system) 多媒体会议系统

MCS (master control system) 播[主]控系统

MCS (multipoint communication service) 多点通信服务

MCU (micro-control unit) 微控制器

MCU (multi-control unit) 多控制单元

MCU (multi-point control unit) 多点控制单元

MD (main distributor) 主配线架

MD (management domain) 管理领域

MD (mini-disc) 迷你光碟[光盘]

MD (motion detection) 移动侦测

MDA (main distribution area) 主配线区

MDB (management database) 管理数据库

MDF (main distributing facility) 主配线设施

MDF (main distribution frame) 主配线架

MDI (medium dependent interface) 媒体相关接口

mean square 均方值

mean thermal transmittance of the wall 外墙平均传热系数

mean time between failure (MTBF) 平均无故障时间

mean time to repair (MTTR) 平均故障修复时间

measure module 测量组件

measurement system 测量系统

M

measuring record 测量记录
mechanical characteristic 机械特性
mechanical damage 机械损坏
mechanical endurance 机械耐久性
mechanical strength 机械强度
mechanical stress 机械应力
mechanical UPS (power) switch 机械式 UPS(电力)开关
MED (medium) 适中,中间(档位)
media access control (MAC) 介质访问控制(子层)
media adapter 媒体适配器
media attachment unit 媒体连接设备
media clock 媒体时钟
media exchange format (MXF) 素材[媒体]交换格式
media packet 媒体包
media switch server (MS) 媒体交换服务器
medium (MED) 媒体
medium access control layer 介质访问控制层
medium dependent interface (MDI) 媒体相关接口
medium interface connector (MIC) 介质接口连接器
medium wave (MW) 中波
meeting room 会议室
megawatt (MW) 兆瓦
MEID (mobile equipment identifier) 移动终端识别号
melt drip 熔滴
melting behaviour 熔融特性
MER (modulation error ratio) 调制误差比[率]
MESH-BN (mesh-bonding network) 网格联结网络
mesh-bonding network (MESH-BN) 网格联结网络
meshed system 网状系统
MET (main earthing terminal) 总接地终端,主接地端子
met by design 通过设计符合
metadata 元数据
metadata class 元数据类
metadata dictionary 元数据字典
metadata element 元数据元素
metadata format 元数据格式
metadata instance 元数据实例
metadata registry database 元数据注册库
metadata schema 元数据框架
metadata service 元数据业务
metal cable trough 金属线槽
metal clip 金属夹
metal hose 金属软管
metal sheath for cable 电缆金属套
metal tube 金属管

M

metal-free 非[无]金属

metal-free fibre optic outdoor cable with central loose tube 带中心松套管的无金属光纤户外电缆

metallic communication cable 金属通信电缆

metallic nitrogen-oxide semiconductor (MNOS) 金属氮氧化物半导体

metallic plumbing 金属管道

meteorological effect 气象效应

metering 计量收费

metering device of energy consumption 能耗计量装置

metering pulse 计费脉冲

metering pulse message (MPM) 计费脉冲消息

metering system of energy consumption 用能计量系统

methane detector 甲烷探测器

metropolitan area network (MAN) 城域网

MF (matched filter) 匹配滤波器

MF (middle frequency) 中频

MFD (mode-field diameter) 模场直径

MFN (multiple frequency network) 多频网

MHP (multimedia home platform) 多媒体家庭平台

MIB (management information base) 管理信息库

MIC (medium interface connector) 介质接口连接器

MIC (microphone) 传声器,话筒

mica paper tape 云母带

MICE classification system MICE 分类系统

MICE environment MICE 环境

micro-control unit (MCU) 微控制器

micro-duct optical fibre 微型光缆

micro-duct 微型吹管

microphone (MIC) 传声器

microphone preamplifier 传声器前置放大器

microphone priority 传声器优先

MicroSD card Micro SD 卡

Microsoft SQL Server 微软关系型数据库管理系统

microwave (MW) 微波

microwave communication 微波通信

microwave detector 微波探测器

microwave double as a detector 微波双鉴探测器

middle frequency (MF) 中频

middleware 中间件

mid-highrise dwelling building 中高层住宅

MIDI (music instrument digital

interface) 音乐设备数字接口

MIDI (musical instrument device interface) 音乐设备接口

milestone 里程碑

MIM (management inhibit message) 管理阻断消息

MIMO (multiple input, multiple output) 多输入多输出

mineral insulated cable 矿物绝缘电缆

mineral insulation noncombustible cable 矿物绝缘类不燃性电缆

mini disc (MD) 迷你光碟

minimum allowable value of energy efficiency 能效比限制,能效最小允许值

minimum bend radius (operating dynamic) 最小弯曲半径(运行动态)

minimum bend radius (operating static) 最小弯曲半径(运行静态)

minimum bending radius (installation) 最小弯曲半径(安装)

minimum coupling loss (MCL) 最小耦合损耗

minimum pulse width (MPW) 最小脉冲宽度

MIRS (multimedia information retrieval system) 多媒体信息检索系统

MIS (management information system) 管理信息系统

missing channel 声道缺失

mixer 调音台

MJPEG (motion joint photographic experts group) 运动静止图像(逐帧)压缩技术

M-LAG (multi-chassis link aggregation group) 跨设备链路聚合组

MLB (main logic board) 主逻辑板

MLD (multicast listener discovery) 组播监听者发现

MM (moving magnet) 动磁式

MMDS (multichannel multipoint distribution services) 多通道多点分配服务

MMF (multimode fiber) 多模光纤

MMF (multimode optical fiber) 多模光纤

MMX (multimedia extensions) 多媒体扩展

MNC (mobile network code) 移动网络代码

MNOS (metallic nitrogen-oxide semiconductor) 金属氮氧化物半导体

MO (managed object) 管理对象

mobile communication 移动通信

mobile communication in-door signal covering system 移动通信室内信号覆盖系统
mobile equipment identifier (MEID) 移动终端识别号
mobile network code (MNC) 移动网络代码
mobile-control 移动控制
MOD (modulation) 调制
modal bandwidth 模式带宽
Modbus Modbus 通信协议
mode-field diameter (MFD) 模场直径
model 86 power supply socket 86 型电源插座
model of mark 型号标志
modem 调制解调器
modern browser 当代[现代]浏览器
modular air conditioning room equipment 模块式空调机房设备
modular connector 模块化连接器
modular system 模块化系统
modularized structure 模块化结构
modulation (MOD) 调制
modulation error ratio (MER) 调制误差比[率]
module 模块,单元模块
module frame 模块框架
module GG45 GG45 模块
module size 模块尺寸
module TERA TERA 模块
moisture 潮湿
monitor 监控(器),监视器
monitoring area 保护面积
monitoring center of energy consumption for buildings 建筑用能监测系统中央控制室
monitoring equipment 监控设备
monitoring radius 保护半径
monitoring system of energy consumption 用能监测系统
mono-fiber cable 单芯光缆
monofilament cable 单芯光缆
mono-mode optical fiber 单模光纤
monophonic signal 单声信号
monophony 单声道技术
mosaic effect 马赛克效果
most significant character (MSC) 最高有效字符
most significant bit (MSB) 最高有效位
motion detection (MD) 移动侦测
motion joint photographic experts group (MJPEG) 运动静止图像(逐帧)压缩技术
motion vector 运动矢量
motorized zoom len 电动变焦镜头
mould 霉菌

mounting 安装
mounting bracket 安装支架
mounting flange 安装法兰
mounting frame 安装框架
mounting support for horizontal installation 水平安装支架
mounting support for vertical installation 垂直安装支架
mouse-proof 防鼠咬
movable equipment 可移动设备
moving magnet (MM) 动磁式
moving picture expert group (MPEG) 运动图像专家组
MP3 (MP3 format) MP3 格式
MP3 (MPEG audio layer-3) MP3 音频压缩技术
MP3 format (MP3) MP3 格式
MPE (multi-protocol encapsulation) 多协议封装
MPEG (moving pictures expert group) 运动图像专家组
MPEG audio layer-3 (MP3) MP3 音频压缩技术
MPEG-1 运动图像专家组规范 1
MPEG-2 运动图像专家组规范 2
MPEG-2 advanced audio coding (AAC) 运动图像专家组规范 2 高级音频编码
MPEG-4 运动图像专家组规范 4
MPEG-7 运动图像专家组规范 7
MPIP (multi-picture-in-picture) 多画面
MPM (metering pulse message) 计费脉冲消息
MPO (multi-fiber push on) MPO 光纤连接器，多芯推进锁闭光纤连接器
MPP (multimedia processing platform) 多媒体处理平台
MPTS (multi-program transport stream) 多节目传输流
MPW (minimum pulse width) 最小脉冲宽度
MS (media switch server) 媒体交换服务器
MS (multiplex section) 复用段
MSA (multiplex section adaptation) 复用段适配
MSB (most significant bit) 最高有效位
MSC (most significant character) 最高有效字符
MSC (MSC anchor) 锚靠 MSC
MSC anchor (MSC) 锚靠 MSC
MSDP (multicast source discovery protocol) 组播源发现协议
MSIF (multimode step-index fiber) 多模阶跃折射率光纤
MSOH (multiplex section overhead) 复用段开销

MSP (multiplex section protection) 复用段保护

MS-SPRING (multiplex section shared protection ring) 复用段共享保护环

MST (multi-metering charging) 复式计次计费

MST (multiplex section termination) 复用段终结

MSTP (multi-service transfer platform) 多业务传输平台

MTBF (mean time between failure) 平均无故障时间

MTCS (multimedia telecommunication conference system) 多媒体电信会议系统

MTP connector MTP 连接器

MT-RJ connector MT-RJ(光纤)连接器

MTTR (mean time to repair) 平均修复时间

MTU (maximum transmission unit) 最大传输单元

MU connector MU(光纤)连接器

multi-control unit (MCU) 多点控制单元

multi-mode fiber (MMF) 多模光纤

multi-user 多用户

multi-agent structure 多自主体结构

multicast 组播

multicast listener discovery (MLD) 组播监听者发现

multicast source discovery protocol (MSDP) 组播源发现协议

multichannel 多通道,多声道

multichannel audio 多声道音频

multichannel audio digital interface (MADI) 多通道音频数字串行接口

multichannel audio stream (MCA-STREAM) 多通道音频传输技术,多声道音频流

multichannel cable 多信道光缆

multichannel digital audio 多声道数字音频

multichannel multipoint distribution services (MMDS) 多通道多点分配服务

multichannel projection 多通道投影

multichannel-bundle cable 多信道束光缆

multi-chassis link aggregation group (M-LAG) 跨设备链路聚合组

multicore and symmetrical pair cable 多芯对绞电缆

multicore and symmetrical pair/quad cable 多芯对称双芯/四芯

M

电缆，对绞或星绞多芯对称电缆

multi-crypt 多密

multi-fiber push on (MPO) MPO光纤连接器，多芯推进锁闭光纤连接器

multifunction controller 多功能控制器

multilevel security proof 多级安全证明

multilingual 多语言

multimedia 多媒体

multimedia application 多媒体应用

multimedia collaborative conference system (MCCS) 多媒体协同会议系统

multimedia communication 多媒体通信

multimedia conferencing system (MCONF) 多媒体会议系统

multimedia extensions (MMX) 多媒体扩展

multimedia home platform (MHP) 多媒体家庭平台

multimedia information recall system 多媒体信息检索系统

multimedia information retrieval system (MIRS) 多媒体信息检索系统

multimedia information system 多媒体信息系统

multimedia processing platform (MPP) 多媒体处理平台

multimedia service 多媒体业务

multimedia service inquiry 多媒体信息查询

multimedia telecommunication conference system (MTCS) 多媒体电信会议系统

multi-metering 多点测量，复式计次

multi-metering charging (MST) 复式计次计费

multimode 10G optical fiber cable 多模万兆光缆

multimode fiber 多模光纤

multimode graded fiber 多模渐变型光纤

multimode modal bandwidth 多模模式带宽

multimode optical fiber (MMF) 多模光纤

multimode optical fiber cable 多模光缆

multimode optical fiber jumper wire 多模光纤跳线

multimode step index fiber (MSIF) 多模阶跃折射率光纤

multi-pair installation cable 多对安装电缆，大对数双绞线

multi-paired cable 多对数电缆
multi-picture-in-picture (MPIP) 多画面
multiple frequency network (MFN) 多频网
multiple function UPS switch 多功能 UPS 开关
multiple input, multiple output (MIMO) 多输入多输出
multiple-loop fire alarm control unit 多路火灾报警控制器
multiplex section (MS) 复用段
multiplex section adaptation (MSA) 复用段适配
multiplex section overhead (MSOH) 复用段开销
multiplex section protection (MSP) 复用段保护
multiplex section shared protection ring (MS-SPRING) 复用段共享保护环
multiplex section termination (MST) 复用段终结
multiplexer (MUX) (多路)复用器
multiplexer and the de-multiplexer 复用器与去复用器
multiplexing unit (MXU) 复用单元
multipoint communication service (MCS) 多点通信服务
multi-point control unit (MCU) 多点控制单元
multi-program transport stream (MPTS) 多节目传输流
multiprotocol BGP (MBGP) 组播协议[多协议]边界网关协议
multi-protocol encapsulation (MPE) 多协议封装
multi-scan monitor 多扫描监视器
multi-service transfer platform (MSTP) 多业务传输平台
multi-stories dwelling building 多层住宅
multi-tandem telephone network 多汇接局电话网
multi-unit cable 多单元线缆
multi-user telecommunications outlet (MUTO) 多用户电信插座
multi-user telecommunications outlet assembly (MUTOA) 多用户电信插座装配
mura 缺陷,斑纹缺陷
music instrument digital interface (MIDI) 音乐设备数字接口
musical instrument device interface (MIDI) 乐器设备接口
multi-screen processor 多屏处理器
MUTO (multi-user telecommunications outlet) 多用户电信插座

M

MUTOA (multi-user telecommunications outlet assembly) 多用户电信插座装配

mutual capacitance 互容抗

MUX (multiplexer) (多路)复用器

MW (medium wave) 中波

MW (megawatt) 兆瓦

MW (microwave) 微波

MWB (manual white balance) 手动白平衡

MXF (media exchange format) 素材[媒体]交换格式

MXU (multiplexing unit) 复用单元

N

N + X redundancy　N + X冗余
N/A (not applicable)　不适用
NAK (negative acknowledgement) 否定应答(信号),否认
NAL (network access license)　进网许可证
name management protocol (NMP) 名字管理协议
name space support (NSS)　名字空间支持
NAP (network access point)　网络接入点
NAPT (network address port translation)　网络地址端口转换
narrowband integrated service digital network (N-ISDN)　窄带综合业务数字网
NAS (network attached storage)　网络连接存储
NAT (network address translator) 网络地址翻译
national destination code (NDC) 国内目的地代码
National Electrical Manufacturers Association (NEMA)　国家电气制造商协会(美国)
National Fire Protection Association (NFPA)　国家防火协会(美国)
national holiday　国家法定假日
national information infrastructure (NII)　国家信息基础设施,信息高速公路
national network congestion (NNC) 国内网拥塞
national roaming　国内漫游
national signaling network (NSN) 国内信令网
national signaling point (NSP)　国内信令点
national standard (NS)　国标,国家标准
National Television Systems Committee (NTSC)　国家电视系统委员会(美国),NTSC制式
national toll number　国内长途电话号码

native signal processing (NSP) 本地信号处理

natural event 自然事件

natural smoke exhausting 自然排烟

navigation 导航

navigation area 导航区

navigation information 导航信息

NC (network computer) 网络计算机

NCC (network color code) 网络色码

NCC (normally closed contact) 常闭触点

NCE (network control engines) 网络控制引擎

NCP (NetWare core protocol) NetWare 核心协议

NCP (network control protocol) 网络控制协议

NCU (network control unit) 网络控制单元

ND (neighbor discovery protocol) 邻居发现协议

NDC (national destination code) 国内目的地代码

NDF (negative dispersion fiber) 负色散光纤

NDIS (network driver interface specification) 网络驱动程序接口规范

NDS (NetWare directory service) NetWare 目录服务

NDSF (non-dispersion shifted fiber) 非色散位移光纤

NE (network element) 网元

near end crosstalk (loss) (NEXT) 近端串扰[音](衰减[损耗])

near field communication (NFC) 近场通信,近距离无线通信技术

near instantaneous companded audio multiplex (NICAM) 准瞬时压扩音复用

near video on demand (NVOD) 准视频点播

NEF (network element function) 网元功能

negative acknowledgement (NAK) 否定应答(信号),否认

negative dispersion fiber (NDF) 负色散光纤

negative resistance effect (NRE) 负电阻效应

negative sequence current (NSC) 负序电流

negative temperature coefficient (NTC) 负温度系数

neighbor discovery protocol (ND) 邻居发现协议

NEL (network element layer) 网

元层

NEMA (National Electrical Manufacturers Association) 国家电气制造商协会(美国)

net weight 净重

NETBEUI (NetBIOS enhanced user interface) NetBIOS 用户扩展接口协议

NetBIOS (network basic input output system) 网络基本输出输入系统

NetBIOS Enhanced User Interface (NETBEUI) NetBIOS 用户扩展接口协议

NETBLT (network block transfer) 网络数据块传送

NetDDE (network dynamic data exchange) 网络动态数据交换

Netscape Netscape 网络网页浏览器

netstream 网络流

NetWare core protocol (NCP) NetWare 核心协议

NetWare directory service (NDS) NetWare 目录服务

network access 进[入]网

network access cabling 网络接入布缆

network access license (NAL) 进网许可证

network access point (NAP) 网络接入点

network adapter 网络适配器

network address port translation (NAPT) 网络地址端口转换

network address translator (NAT) 网络地址翻译

network architecture 网络体系结构

network attached storage (NAS) 网络连接存储

network automation 网络自动化

network bandwidth 网络带宽

network basic input output system (NetBIOS) 网络基本输出输入系统

network block transfer (NETBLT) 网络数据块传送

network bridge 网桥

network capacity 网络容量[能力]

network card 网卡

network card driver 网卡驱动程序

network clock 网络时钟

network color code (NCC) 网络色码

network computer (NC) 网络计算机

network control engine (NCE) 网络控制引擎

network control protocol (NCP) 网络控制协议

network control unit (NCU) 网络控制单元

network conversion interface 网络转换接口

network dynamic data exchange (DDE) service 网络动态数据交换服务

network driver interface specification (NDIS) 网络驱动程序接口规范

network element (NE) 网元

network element function (NEF) 网元功能

network element layer (NEL) 网元层

network file system (NFS) 网络文件系统

network gateway 网关

network ID (network identifier) 网络标识符

network identification (NID) 网络识别码

network identifier (network ID) 网络标识符

network information center (NIC) 网络信息中心

network information service (NIS) 网络信息服务

network information system (NIS) 网络信息系统

network information table (NIT) 网络信息表

network integration engines (NIE) 网络集成引擎

network interface (NI) 网络接口

network interface card (NIC) 网络接口卡,网卡

network interface definition language (NIDL) 网络接口定义语言

network interface layer (NIL) 网络接口层

network interface unit (NIU) 网络接口部件

network junction 网络结[节]点

network layer 网络层

network management 网络管理

network management entity (NME) 网络管理实体

network management framework (NMF) 网络管理框架

network management function (NMF) 网络管理功能

network management gateway (NMG) 网络管理网关

network management information system (NMIS) 网络管理信息系统

network management interface

(NMI) 网络管理接口

network management layer (NML) 网络管理层

network management model (NMM) 网络管理模型

network management plan (NMP) 网络管理计划

network management protocol (NMP) 网络管理协议

network management subsystem (NMS) 网络管理子系统

network management system (NMS) 网络管理系统

network management vector transport (NMVT) 网络管理向量传输

network manager 网络管理器[员]

network-network interface (NNI) 网络-网络接口

network news transfer protocol (NNTP) 网络新闻传输协议

network node 网络节点

network node interface (NNI) 网络节点接口

network of antennas 天线网络

network operating system (NOS) 网络操作系统

network printer 网络打印机

network record unit (NRU) 网络记录[录像]单元

network security system 网络安全系统

network service access point (NSAP) 网络服务接入点

network service protocol (NSP) 网络服务协议

network service provider (NSP) 网络服务提供商

network subsystem 网络子系统

network surveillance system (NSS) 网络监视系统

network synthesis of multimedia system 多媒体系统网络集成

network system 网络系统

network television 网络电视

network terminal adapter (NTA) 网络终端适配器

network termination (NT) 网络终端

network time protocol (NTP) 网络时间协议

network topology 网络拓扑

network video recorder (NVR) 网络视频录像机

network virtual terminal (NVT) 网络虚拟终端

network voice protocol (NVP) 网络语音协议

networking cable 网络线缆

network-to-network interface (NNI)

网络-网络接口

neural network control (NNC) 神经网络控制

neuron chip 神经元芯片

neutral conductor 中性导体

neutral line 中性线

NEXT (near end crosstalk loss) 近端串扰[音],近端串扰衰减(损耗)

next generation broadcasting (NGB) 下一代广播电视网

next generation broadcasting-html (NGB-H) 基于 HTML 的下一代广播电视网中间件

next generation broadcasting-Java (NGB-J) 基于 Java 的下一代广播电视网中间件

next generation Internet (NGI) 下一代因特网

next generation network (NGN) 下一代网络

next hop resolution protocol (NHRP) 下一跳解析协议

NEXT worst pair 近端串扰[音],近端串扰衰减(损耗)

NFC (near field communication) 近场通信,近距离无线通信技术

NFPA (National Fire Protection Association) 国家防火协会(美国)

NFS (network file system) 网络文件系统

NGB (next generation broadcasting) 下一代广播电视网

NGB-H (next generation broadcasting-html) 基于 HTML 的下一代广播电视网中间件

NGB-J (next generation broadcasting-Java) 基于 Java 的下一代广播电视网中间件

NGI (next generation Internet) 下一代因特网

NGN (next generation network) 下一代网络

NH/LSZH (low smoke zero halogen) 耐火低烟无卤

NHRP (next hop resolution protocol) 下一跳解析协议

NI (network interface) 网络接口

NIC (network information center) 网络信息中心

NIC (network interface card) 网络接口卡,网卡

NICAM (near instantaneous companded audio multiplex) 准瞬时压扩音复用

NID (network identification) 网络识别码

NIDL (network interface definition language) 网络接口定义语言

NIE (network integration engines) 网络集成引擎

NII (national information infrastructure) 国家信息基础设施,信息高速公路

NIL (network interface layer) 网络接口层

NIM (non-intrusive monitoring) 非介入监控

NIS (network information service) 网络信息服务

NIS (network information system) 网络信息系统

NIS (number information service) 号码信息服务

N-ISDN (narrowband integrated service digital network) 窄带综合业务数字网

NIT (network information table) 网络信息表

NIU (network interface unit) 网络接口部件

NLP (normal link pulse) 正常链路脉冲

NMC (non-metallic component) 非金属部件

NME (network management entity) 网络管理实体

NMF (network management framework) 网络管理框架

NMF (network management function) 网络管理功能

NMG (network management gateway) 网络管理网关

NMI (network management interface) 网络管理接口

NMI (non-maskable interrupt) 非屏蔽中断

NMIS (network management information system) 网络管理信息系统

NML (network management layer) 网络管理层

NMM (node message memory) 节点消息存储器

NMP (name management protocol) 名字管理协议

NMP (network management plan) 网络管理计划

NMP (network management protocol) 网络管理协议

NMS (network management subsystem) 网络管理子系统

NMS (network management system) 网络管理系统

NMVT (network management vector transport) 网络管理向量传输

NNC (national network congestion) 国内网拥塞

NNC (neural network control) 神

经网络控制

NNI (network-network interface) 网络-网络接口

NNI (network node interface) 网络节点接口

NNI (network-to-network interface) 网络-网络接口

NNTP (network news transfer protocol) 网络新闻传输协议

no action temperature 不动作温度

no burner 不燃烧体

no corrosive gas 无腐蚀性气体

no load current 空载电流

NOC (normally open contact) 常开触点

nodal processing delay 节点处理时延

node identifier 节点标识符

node message memory (NMM) 节点消息存储器

node of electric circuit 电路节点

node to node message 节点间消息

noise 噪声

noise gate 噪声门

noise level 噪声电平

noise reduction 消声,降噪

noise suppressor 噪声抑制器

nominal buffer diameter 标称缓冲区直径

nominal characteristic impedance 标称特性阻抗

nominal cladding diameter 标称包层直径

nominal diameter 标称直径

nominal impedance 标称阻抗

nominal output power (NOP) 额定输出功率

nominal value 标称值

nominal velocity of propagation (NVP) 标称传播[相对光速]速度

nominal voltage 标称电压

non-balanced interface 非平衡接口

non-blocking 非阻塞

non-circularity 不圆度

non-crystal 非晶体

non-detachable power supply cord 不可拆卸的电源软线

non-dispersion shifted fiber (NDSF) 非色散位移光纤

non-fire power supply 非消防电源

non-highly sensitive detector 非高灵敏型探测器

non-intrusive monitoring (NIM) 非介入监控

non-linear component 非线性分量

non-linear distortion 非线性失真

non-linear load 非线性负载
non-maskable interrupt (NMI) 非屏蔽中断
non-metal optical fiber cable 无金属光缆
non-metallic component (NMC) 非金属部件
non-metalware 非金属件
non-packet terminal (NPT) 非分组终端
non-redundant circuit 无冗余电路
non-return-to-zero (NRZ) 不归零制
non-revertive path switching 非恢复式通道倒换
non-traditional water source 非传统水源
non-voice service 非电话[语音]业务
non-voice terminal 非话终端
non-volatile random access memory (NVRAM) 非挥发性随机读写存储器
non-zero dispersion fiber (NZDF) 非零色散光纤
non-zero dispersion shifted fiber (NZDSF) 非零色散位移光纤
NOP (nominal output power) 额定输出功率
normal lighting 正常照明
normal link pulse (NLP) 正常链路脉冲
normal load 正常负载
normal play time (NPT) 标准播放时间
normal response mode (NRM) 正常响应方式
normalized response mode (NRM) 归一化响应模式
normalized signal/noise ratio 归一化信噪比
normally closed contact (NCC) 常闭触点
normally open contact (NOC) 常开触点
NOS (network operating system) 网络操作系统
not applicable (N/A) 不适用
NP (number portability) 号码可携性，号码移动性
NPT (non-packet terminal) 非分组终端
NPT (normal play time) 标准播放时间
NRE (negative resistance effect) 负电阻效应
NRM (normal response mode) 正常响应方式
NRM (normalized response mode) 归一化响应模式

NRU (network record unit) 网络记录[录像]单元

NRZ (non-return-to-zero) 不归零制

NS (national standard) 国标,国家标准

NSAP (network service access point) 网络层服务访问点,网络业务接入点

NSC (negative sequence current) 负序电流

NSN (national signaling network) 国内信令网

NSP (national signaling point) 国内信令点

NSP (native signal processing) 本地信号处理

NSP (network service protocol) 网络服务协议

NSP (network service provider) 网络服务提供商

NSS (name space support) 名字空间支持

NSS (network surveillance system) 网络监视系统

NT (network termination) 网络终端

NTA (network terminal adapter) 网络终端适配器

NTC (negative temperature coefficient) 负温度系数

NTP (network time protocol) 网络时间协议

NTSC (National Television Systems Committee) 国家电视系统委员会(美国),NTSC 制式

nuclear user data 核心用户数据

number allocation 放号

number analysis table 号码分析表

number information service (NIS) 号码信息服务

number of pairs 对数

number of windings 环绕圈数

number portability (NP) 号码可携性,号码移动性

number segment 号码段

number storing key 号码存储键

numerical aperture 数值孔径

nurse call 护士呼叫

nurse call system 护士呼叫系统

NVOD (near video on demand) 准视频点播

NVP (network voice protocol) 网络语音协议

NVP (nominal velocity of propagation) 标称传播[相对光速]速度

NVR (network video recorder) 网络视频录像机

NVRAM (non-volatile random access memory) 非挥发性随机读

写存储器

NVT (network virtual terminal) 网络虚拟终端

Nyquist 奈奎斯特(美国物理学家)

NZDF (non-zero dispersion fiber) 非零色散光纤

NZDSF (non-zero dispersion shifted fiber) 非零色散位移光纤

O

O/E (optical to electrical converter) 光电转换器

OA (office automation) 办公自动化

OA (open architecture) 开放式体系结构,开放架构

OA (optical amplifier) 光放大器

OAA (open application architecture) 开放式应用体系结构

OADM (optical add/drop multiplexer) 光波分插复用器

OAN (optical access network) 光纤接入网

OAS (office automation system) 办公自动化系统

OB van (outside broadcast van) 转播车

OBD (optical branching device) 光分支装置

object 目标,实体,对象

object linking and embedding (OLE) 对象链接与嵌入

objective 可观的

oblique cutting wall faceplate 斜插墙面面板

obstructed area 障碍区域

OC (open circuit) 断开电路,开路

OC (optical coupler) 光耦合器

OC-1 (optical carrier level 1) 第一级光载波

OCA (optical channel analyzer) 光信道分析仪

occupational health 职业健康

OCDP (optical coherence domain polarimetry) 光相干域偏振(检测技术)

OCDR (optical coherence domain reflectometer) 光相干域反射仪

OCF (optical cable facility) 光纤设施

OCH (optical channel) 光通道[信道]

OCL (optical channel layer) 光信道层

OCL (output capacitorless) 无输出电容的功率放大器(OCL 电路)

O

OC-N (optical carrier level N) 第N级光载波

OCR (optical character recognition) 光学字符识别

OCS (optical character scanner) 光学字符扫描器

OCS (optical coherent system) 相干光系统

octave 倍频程

octave band sound pressure level 倍频带声压级

OCTS (optical cable transmission system) 光缆传输系统

ODBC (open database connectivity) 开放数据库互联

ODF (optical distribution frame) 光配线架

ODF (optical fiber distribution frame) 光纤配线架

ODI (open data link interface) 开放数据链路接口

ODN (optical distribution network) 光分配网络

ODN (optical distribution node) 光分配节点

ODS (on-demand service) 即时业务

ODSI (optical domain service interconnect) 光域业务互联

ODT (optical data transmission) 光数据传输

ODU (optical demultiplexing unit) 光分波单元,光分波器

OE EQP (opto-electronic equipment) 光电设备

OEIC (opto-electronic integrated circuit) 光电集成电路

OEID (opto-electronic integrated device) 光电集成器件

OF (optical fibre) 光纤

OFA (optical fiber amplifier) 光纤放大器

OFCC (optical fiber cable component) 光纤电缆组件

OFDM (orthogonal frequency division multiplexing) 正交频分复用

OFDR (optical frequency domain reflectometer) 光频域反射计

OFE (optical fiber equalizer) 光纤均衡器

off line box 过线盒

office automation (OA) 办公自动化

office automation system (OAS) 办公自动化系统

office building 办公楼,办公建筑

offline simulation 离线仿真

off-setted sample 偏移后样本

off-the-film-metering 焦平面测光

OFL (overfilled launch) 满注入

OFL-BW (overfilled launch bandwidth) 过满注入带宽

OFM (optical frequency modulation) 光频调制

OFP (optical fiber path) 光纤通道

OFS (optical fiber sensor) 光纤传感器

often open fire door 常开防火门

OggVorbis OggVorbis 音频编码格式

OGSA (open grid services architecture) 开放式网格服务架构

OHA (overhead access) 开销接入

ohmmeter 欧姆表

OIC (optical integrated circuit) 光集成电路

oil burning boiler 燃油锅炉

oil mist 油雾

oil resistant 抗油污(的)

oil tank 石油储罐

OIU (optical transmit unit) 光发射单元

OL (optical line) 光线路

OLA (optical line amplifier) 光线路放大器

OLE (object linking and embedding) 对象链接与嵌入

OLED (organic light-emitting diode) 有机发光二极管

OLT (optical line terminal) 光线路终端

OLTS (optical loss test set) 光损耗测试仪,光损耗测试

OMP (operation management procedure) 运行管理程序

OMS (outage management system) 停电管理系统

OMU (optical multiplexer unit) 光合波板,光复用单元

on-demand service (ODS) 即时业务

one cartoon system 一卡通系统

one touch ready 一键到位

one way cable TV system 单向电缆电视系统

one way stop return valve 单向止回阀

one-layer STP 单层屏蔽双绞线

on-hook 挂机(电话)

online alarm system 联机报警系统

on-screen display (OSD) 屏幕显示,屏幕菜单式调节方式

on-site organization 现场组织机构

on-site test 现场测试

ONU (optical network unit) 光网络单元

ONVIF (Open Network Video Interface Forum) 开放型网络视频接口

论坛

OOF (out of frame) 帧失步

OPA (optical preamplifier) 光前置放大板

opacity of smoke 烟的阻光度

open 开口，开启

open application architecture (OAA) 开放式应用体系结构

open architecture (OA) 开放式体系结构，开放结构

open circuit (OC) 断开电路，开路

open construction 开放结构

open database connectivity (ODBC) 开放数据库互联

open data-link interface (ODI) 开放数据链路接口

open fire operation 明火作业

open grid services architecture (OGSA) 开放式网格服务架构

Open Network Video Interface Forum (ONVIF) 开放型网络视频接口论坛

open office cabling 开放式办公室布线

open office cabling system 开放型办公室布线系统

open pluggable specification (OPS) 开放式可插拔规范

open service architecture (OSA) 开放的业务结构

open shortest path first (OSPF) 开放式最短路径优先(算法)

open staircase 敞开楼梯间

open stairway 敞开楼梯

open system interconnection (OSI) 开放系统互联

open system interconnection reference model (OSI/RM) 开放系统互联参考模型

open system interconnect (OSI) 开放系统互联

open system interconnection model (OSI model) 开放式系统互联模型

opening angle 开启角度

operating distance 操作距离

operating dynamic 运行动态

operating static 运行静态，静态操作

operating voltage 工作电压

operating wavelength 工作波长

operation 操作，运算，运行

operation against rule 违章操作

operation and command center 运行指挥中心

operation and maintenance 操作和维护

operation and maintenance department of the radio broadcasting center 广播中心运维单位

operation and maintenance manual 操作与维护手册

operation interface 操作界面

operation management procedure (OMP) 运行管理程序

operation management seat 操作管理席位

operation manual 操作手册

operation support system (OSS) 运营支撑系统

operational life 工作寿命

operational log 运行记录

operational requirement 操作要求

operational test for automatic fire alarm system 火灾自动报警系统效用试验

operational test for automatic sprinkler system 自动喷水系统工作试验

operational test for inert gas system 惰性气体系统效用试验

operational test for water fire-extinguishing system 水灭火系统效用试验

operations, maintenance and administration part 运行、维护和管理部分

operator 操作符,操作,操作员

operator of CATV network 有线电视网络运营机构

OPLC (optical fiber composite low-voltage cable) 光纤复合低压电缆

OPM (optical power meter) 光功率计

OPS (open pluggable specification) 开放式可插拔规范

optical access network (OAN) 光接入网

optical add/drop multiplexer (OADM) 光波分插复用器

optical alarming mode 光告警方式

optical amplifier (OA) 光放大器

optical attenuation 光衰减

optical attenuator 光衰减器

optical branching device (OBD) 光分支装置

optical broadcasting system with narrow band overlay 窄播光插入系统

optical cable 光缆

optical cable connector 光缆连接器

optical cable cross-connecting cabinet 光缆交接箱

optical cable end joint 光缆成端接头

optical cable facility (OCF) 光缆设施

optical cable transmission system

(OCTS) 光缆传输系统

optical carrier level 1 (OC-1) 第一级光载波

optical carrier level N (OC-N) 第N级光载波

optical channel (OCH) 光通道[信道]

optical channel analyzer (OCA) 光信道分析仪

optical channel layer (OCL) 光信道层

optical character recognition (OCR) 光学字符识别

optical character scanner (OCS) 光学字符扫描器

optical circulator 光环行器

optical coherence domain polarimetry (OCDP) 光相干域偏振(检测技术)

optical coherence domain reflectometer (OCDR) 光相干域反射仪

optical coherent system (OCS) 相干光系统

optical coupler (OC) 光耦合器

optical cross-connect (OXC) 光交叉连接(光互联)

optical crosstalk at individual channel output port 单个光通道输出端口的光串扰

optical data transmission (ODT) 光纤数据传输

optical demultiplexing unit (ODU) 光分波单元,光分波器

optical density of smoke 烟的光密度

optical digital cross connect 光数字交叉连接

optical distribution frame (ODF) 光配线架

optical distribution network (ODN) 光分配网络

optical distribution node (ODN) 光分配节点

optical divider 光分路器

optical domain service interconnect (ODSI) 光域业务互联

optical fiber amplifier (OFA) 光放大器

optical fiber amplifier used in CATV system 有线电视系统光纤放大器

optical fiber cable 光纤光缆

optical fiber cable communication system 光缆通信系统

optical fiber cable component (OFCC) 光缆元件

optical fiber channel 光纤信道

optical fiber cluster 光导纤维束

optical fiber composite low-voltage cable (OPLC) 光纤复合低压

电缆

optical fiber connecting and distributing unit 光纤配线单元

optical fiber connector 光纤连接器

optical fiber coupler 光纤耦合器

optical fiber delay line 光纤延时线

optical fiber distribution frame (ODF) 光纤配线架

optical fiber equalizer (OFE) 光纤均衡器

optical fiber fusion 光纤熔接

optical fiber jumper 跳纤

optical fiber LAN (optical fiber local area network) 光纤局域网

optical fiber local area network (optical fiber LAN) 光纤局域网

optical fiber mechanical splice 光纤机械式接续器,光纤机械接头

optical fiber patch cord 光纤跳线

optical fiber patch panel 光纤配线架

optical fiber path (OFP) 光纤通道

optical fiber pigtail 光纤尾纤

optical fiber sensor (OFS) 光纤传感器

optical fiber splicing 光纤接续

optical fiber splitter 光分路器

optical fiber terminal box 光纤终端盒

optical fiber transmission 光纤传输

optical fiber transmission system 光纤传输系统

optical fibre (OF) 光纤

optical fibre adapter 光纤适配器

optical fibre cable 光纤线缆

optical fibre cable (or optical cable) 光纤线缆(或光缆)

optical fibre duplex adapter 光纤双工适配器

optical fibre duplex connector 光纤双工连接器

optical fibre pair 光纤对

optical fibre perform 光纤预制棒

optical fibre polarity 光纤极性

optical fibre ribbon cable 带状光缆

optical fibre type 光纤类型

optical filter 光滤波器

optical fixed attenuator 光纤固定衰减器

optical flame fire detector 感光火灾探测器

optical frequency division multiple address (FDMA) 光频分多址

optical frequency division multiplexing 光频分复用

optical frequency domain reflectometer (OFDR) 光频域反射计
optical frequency modulation (OFM) 光频调制
optical image quality 光学图像质量
optical integrated circuit (OIC) 光集成电路
optical interconnection 光互连
optical isolator 光隔离器
optical lens 光学镜头
optical line (OL) 光线路
optical line amplifier (OLA) 光线路放大器
optical line terminal (OLT) 光线路终端
optical link 光链路
optical loss test set (OLTS) 光损耗测试仪,光损耗测试
optical modem 光调制解调器
optical modulator 光调制器
optical multimeter 光多用表
optical multiplexer 光复用器
optical multiplexer unit (OMU) 光复用单元,光合波板
optical network unit (ONU) 光网络单元
optical node 光节点
optical passive device 光无源器件
optical power difference 光功率差
optical power machine 光动力机器,光能发电机
optical power meter (OPM) 光功率计
optical preamplifier (OPA) 光前置放大板
optical receiver 光接收机
optical receiving component 光接收器件
optical repeater 光中继器
optical resistor 光敏电阻
optical router 光纤路由器
optical scanner 光学扫描器
optical signal 光信号
optical splitter 光分路器
optical switch 光开关
optical telecommunication outlet (OTO) 光通信插座,光纤信息插座
optical time division multiplexing (OTDM) 光时分复用
optical time-domain reflectometer (OTDR) 光时域反射计
optical to electrical connection 光电连接
optical to electrical converter (O/E) 光电转换器
optical transceiver 光收发模块
optical transceiver board (OTB) 光收发板
optical transmission 光传输

optical transmission network (OTN) 光传输网络

optical transmit unit (OIU) 光发送单元

optical transmitter 光发送机

optical transmitting component 光发送器件

optical transmitting LED 光发射二极管

optical wavelength 光波长

optical zoom 光学变焦

optics 光学

optional connection 可选连接

opto-electrical transceiver 光电收发器

opto-electronic device 光电器件

opto-electronic equipment (OE EQP) 光电设备

opto-electronic integrated circuit (OEIC) 光电集成电路

opto-electronic integrated device (OEID) 光电集成器件

opto-electronic process 光电处理

orderly 有序

ordinary broadcast 普通广播

ordinary clock 普通时钟

ordinary electric energy meter 普通电能表

ordinary telephone 普通电话

organic glass label frame 有机玻璃标签框

organic light-emitting diode (OLED) 有机发光二极管

organic material 有机物质

organization and management structure of field project 现场项目组织管理结构

organizationally unique identifier (OUI) 唯一组织标识符

originating party 发起方

orthogonal frequency division multiplexing (OFDM) 正交频分复用

OSA (open service architecture) 开放的业务结构

OSC (oscillator) 振荡器

oscillator (OSC) 振荡器

OSD (on-screen display) 屏幕显示,屏幕菜单式调节方式

OSI (open system interconnection) 开放系统互联

OSI (open system interconnect) 开放系统互联

OSI model (open system interconnection model) 开放式系统互联模型

OSI/RM (open system interconnection reference model) 开放系统互联参考模型

OSPF (open shortest path first) 开

放式最短路径优先(算法)

OSS (operation support system) 运营支撑系统

OTB (optical transceiver board) 光收发板

OTDM (optical time division multiplexing) 光时分复用

OTDR (optical time-domain reflectometer) 光时域反射计

OTL (output transformerless) 推挽式无输出变压器功率放大电路(OTL 电路)

OTN (optical transmission network) 光传输网络

OTO (optical telecommunication outlet) 光通信插座,光纤信息插座

OUI (organizationally unique identifier) 唯一组织标识符

out of frame (OOF) 帧失步

outage management system (OMS) 停电管理系统

outcoming media stream 出口媒体流

outdoor base station 室外基站

outdoor bushing 户外套管

outdoor cable 室外缆线

outdoor enclosure 户外机壳

outdoor equipment cabinet 室外设备箱

outdoor external insulation 户外外绝缘

outdoor LED display screen 户外发光二极管[LED]显示屏

outdoor optical fiber 室外光缆

outdoor optical fibre cable 室外光纤线缆

outdoor single-mode optical fiber cable 室外单模光缆

outdoor unit 室外机,门口机

outdoor-immersed bushing 户外浸入式套管

outdoor-indoor bushing 户外-户内套管

outer diameter 外径

outer diameter of backbone cable 主干线缆外径

outer sheath 外护套

outfire 灭火

outlet 插座

outlet level 输出电平

outlet, telecommunication 出口(电信)

outlet/connector, telecommunication 出口连接器(电信),插座连接器(电信)

out-of-band 带外

out-of-band emission 带外发射

out-of-service 非服务状态

output active power 输出有功功率

output apparent power 输出视在功率
output apparent power-reference non-linear loading 基准非线性负载时的输出视在功率
output capacitorless (OCL) 无输出电容的功率放大器(OCL 电路)
output current 输出电流
output frequency tolerance 输出频率允差
output impedance 输出阻抗
output order 输出顺序
output over current 输出过电流
output transformerless (OTL) 推挽式无输出变压器功率放大电路(OTL 电路)
output voltage 输出电压
output voltage tolerance 输出电压允差
outside broadcast van (OB van) 转播车
outside shading coefficient 外遮阳系数
outside telephone 外线电话
ovality 椭圆度
over load capability 过载能力
over-/under-voltage 过欠压
overall copper braid 编织铜网总屏蔽层
overall foil and copper braid 铝箔和丝网总屏蔽
overall foil screen 铝箔总屏蔽
overall screen 总屏蔽层
overall screen foil 铝箔总屏蔽
overdesign 超安全标准设计
overfilled launch (OFL) 满注入
overfilled launch bandwidth (OFL-BW) 过满注入带宽
overhead access (OHA) 开销接入
overhead communication cable 架空通信电缆
overload protection 过载保护
override 强插
oversampling 过采样
overscan 过扫描
overseas training charge 海外培训费用
overshoot 过冲
over-voltage 过电压
over-voltage protection (OVP) 过压保护
OVP (over-voltage protection) 过压保护
OXC (optical cross-connect) 光交叉连接(光互联)
oxygen index 氧指数

P

P frame P帧,帧间预测编码帧

P & T motors-controls P & T云台控制

P/S (parallel/serial conversion) 并串变换

P2MP (point to multipoint) 点到多点

P2P (peer-to-peer) 对等

P2P (point-to-point) 点到点

P2P network (peer-to-peer network) 对等网络

PA (parking automation) 停车管理系统

PA (process automation) 过程自动化

PA (public address) 公共广播

PABX (private automatic branch exchange) 专用自动交换分机

PACCH (packet association control channel) 分组随路控制信道

packet access grant channel (PAGCH) 分组接入应答信道

packet assembler-disassembler (PAD) 分组[分包]装配拆卸器

packet assembling /de-assembling (PAD) 分组装拆,包组装拆装

packet assembly and disassembly device (PAD) 分组拆装设备

packet association control channel (PACCH) 分组随路控制信道

packet broadcast control channel (PBCCH) 分组广播控制信道

packet control unit (PCU) 分组控制单元

packet data channel (PDCH) 分组数据信道

packet data network (PDN) 分组数据网络

packet data protocol (PDP) 分组数据协议

packet data support node (PDSN) 分组数据支持节点

packet data traffic channel (PDTCH) 分组业务数据信道,分组数据业务信道

packet delay 分组延迟

packet digital terminal equipment 分组数字终端设备
packet handler (PH) 分组处理器
packet identifier (PID) 包标识符
packet layer protocol (PLP) 分组层协议
packet paging channel (PPCH) 分组寻呼信道
packet random access channel (PRACH) 分组随机接入信道
packet retransmission (PR) 分组转发
packet switched data network (PSDN) 公共分组数据网络
packet switched public data network (PSPDN) 分组交换公用数据网
packet switching (PS) 分组[分包]交换
packet switching exchange (PSE) 分组交换机
packet switching protocol 分组交换协议
packet terminal (PT) 分组终端
packet time 包时间
packet timing advanced control channel (PTCCH) 分组定时提前控制信道
packet TMSI (P-TMSI) 分组临时移动用户识别码,分组 TMSI
packetized elementary stream (PES) 打包的基本码流
packing list 装箱单
PAD (packet assembler-disassembler) 分组[分包]装配拆卸器
PAD (packet assembling /de-assembling) 分组装拆,包组装拆
PAD (packet assembly and disassembly device) 分组拆装设备
PAD (program associated data) 节目相关数据
PAGCH (packet access grant channel) 分组接入应答信道
paging 寻呼
pair 线对
pair seperator (cross) (十字)线对隔离
PAL (phase alteration by line) 逐行倒相,帕尔制
PAM (pluggable authentication module) 可插入认证模块
PAM (pulse amplitude modulation) 脉(冲)幅(度)调制
pan/tilt/zoom (PTZ) 云台全方位控制
pane 窗格
panic button 紧急按钮
PANID (personal area network ID) 个域网标识符
PAP (password authentication

protocol) 密码验证[认证]协议
parallel earthing conductor (PEC) 平行接地导体
parallel optic port 并行光学端口
parallel plier 平行钳
parallel redundant UPS 并联冗余 UPS
parallel UPS 并联 UPS
parallel/serial conversion (P/S) 并串变换
parametric equalizer 参数均衡器
parental control 家长控制
park equipment 停车场设备
parking automation (PA) 停车管理系统
parking lot 停车场
parking lot (library) management system 停车场(库)管理系统
parking lot management system (PLMS) 停车库(场)管理系统
parking management system (PMS) 停车管理系统
parse 解析
partial grant 部分授权
partial packet discard (PPD) 部分分组丢弃
partial parallel UPS 局部并联 UPS
partition 分区,划分
partition type 分区类型
PAS (personal access system) 个人接入系统
PAS (public address system) 公共广播系统
pass box 过路箱[盒],传递箱
pass criteria 合格判据
passive 无源
passive 3D 被动 3D
passive component 无源部件
passive element 无源电路
passive infrared (PIR) 被动红外
passive infrared detector (PID) 被动红外探测器
passive network 无源网络
passive optical network (PON) 无源光网络
passive optical splitter (POS) 无源光纤分光器
passive section 无源部件
passive speaker 无源音箱
passive terminal / active terminal 无源终端与有源终端
password authentication protocol (PAP) 密码验证[认证]协议
PAT (program association table) 节目关联表
patch cord 快接跳线
patch jumper 跳线
patch panel (PP) 配线架
patent right 专利权

pathway 路径

pathway system 通道系统

patient monitoring system (PMS) 病人监护系统

patient room 病房

patrol system 巡更系统

pay per channel (PPC) 按频道付费

pay per event 按事件付费

pay per view (PPV) 按次付费

payload 有效负荷

payload pointer 负荷指针,有效载荷指针

payload type (PT) 有效载荷类型

pay-TV 付费电视

PBCCH (packet broadcast control channel) 分组广播控制信道

PBX (private branch exchange) 程控用户交换机

PC (physical contact) 物理接触

PC (polycarbonate) 聚碳酸酯

PCA (policy certification authorities) 认证管理机构

PCC (programmable computer controller) 可编程计算机控制器

PCF (point coordination function) 点协调功能

PCI (peripheral component interconnect) 外围部件互连

PCI mezzanine card (PMC) PCI夹层卡

PCM (pulse code modulation) 脉冲编码调制

PCN (personal communication network) 个人[专用]通信网络

PCnet conference 电脑网络会议

PCR (peak cell rate) 峰值信元速率

PCR (program clock reference) 节目时钟基准

PCS (personal communication service) 个人通信服务

PCS (personal communication system) 个人通信系统

PCU (packet control unit) 分组控制单元

PDA (personal digital assistant) 个人数字助理

PDC (personal digital cellular telecommunication system) 个人数字蜂窝通信系统

PDC (personal digital communication) 个人数字通信

PDCH (packet data channel) 分组数据信道

PDF (portable document format) 便携式文档格式

PDH (plesiochronous digital hierarchy) 准同步数字系列

PDN (packet data network) 分组数据网络
PDN (private data network) 专用数据网络
PDN (public data network) 公共数据网络
PDN (public network) 公共[公用]网络
PDP (packet data protocol) 分组数据协议
PDP (plasma display panel) 等离子显示器,电浆显示板
PDP (power distribution panel) 配电盘
PDR (polarization diversity receiver) 偏振[极化]分集接收机
PDR (program dynamic range) 节目(音频)动态范围
PDS (personal digital system) 个人数字系统
PDS (premises distribution system) 综合布线系统
PDSN (packet data support node) 分组数据支持节点
PDTCH (packet data traffic channel) 分组业务数据信道,分组数据业务信道
PE (polyethylene) 聚乙烯
PE (protective earthing conductor) 保护性接地导体
peak cell rate (PCR) 峰值信元速率
peak factor 峰值因数
peak signal to noise ratio (PSNR) 峰值信噪比
peak voltage variation 峰值电压变化
PEC (parallel earthing conductor) 平行接地导体
peer-to-peer (P2P) 对等
peer-to-peer network (P2P network) 对等网络
PEM (processing element memory) 局部存储器,处理元件存储器
PEN conductor (protective earth and neutral conductor) PEN导体
penetration depth 渗透[穿透]深度
per channel input power 每通道输入功率
per channel optical signal-to-noise ratio 单路光通道光信噪比
per channel output power 每通道输出功率
performance limit 性能极限值
performance parameter 性能参数
performance test instrument 性能测试仪器
performance test model (PTM) 性能测试模型

performance tester 性能测试仪
performance testing 性能测试
perimeter 周界
perimeter precaution 周界防范
periodic output voltage variation 输出电压的周期性变化
peripheral component interconnect (PCI) 外围部件互连
peripheral unit (PU) 外围单元[设备]
permanent link (PL) 永久链路
permanent project 永久性工程
permanent virtual circuit (PVC) 永久虚拟线路
permanent virtual connection (PVC) 永久虚连接
permanent wall 永久墙
permanently connection equipment 永久连接式设备
permeability 磁导率
permit 许可证
permitted frequency deviation of optical input interface 光输入口允许频偏
personal access system (PAS) 个人接入系统
personal area network ID (PANID) 个域网标识符
personal authentication key 个人身份验证密钥
personal communication network (PCN) 个人[专用]通信网络
personal communication service (PCS) 个人通信服务
personal communication system (PCS) 个人通信系统
personal digital assistant (PDA) 个人数字助理
personal digital cellular (PDC) telecommunication system 个人数字蜂窝通信系统
personal digital communication (PDC) 个人数字通信
personal digital system (PDS) 个人数字系统
personal handphone system (PHS) 个人手持电话系统，双向无绳电话系统(日本)
personal identification number (PIN) 个人身份识别码
personal number (PN) 个人号码
personal telecommunication number (PTN) 个人通信号码
personal well-being 个人健康
personnel protection 人力防范
PES (packetized elementary stream) 打包的基本码流
PES packet header PES包头
PES stream PES码流
PFA (predictive failure analysis)

故障预警分析

PFC (power factor correction) 功率因数校正

PGP (pretty good privacy) 优良密保(协议)

PH (packet handler) 分组处理器

phase alternating by line (PAL) 逐行倒相

phase angle 相位角

phase error 相位误差

phase shift keying (PSK) 相移键控

phone panel 语音配线架

phosphatize 磷化

phosphor bronze alloy 铜磷合金，磷青铜

photocoupler 光(电)耦合器

photodetector 光探测器

photoelectric 光电的

photoelectric beam detector 光束遮断式感应器

photoelectric converter 光电转换器

photoelectric encoder 光电编码器

photoelectric imaging technology 光电成像技术

photoelectric integrated machine 光电整机

photoelectric isolation 光电隔离

photoelectric sensor 光电传感器

photoelectric switch 光电开关

photonic cross connect (PXC) 全光交叉连接

photovoltaic cables for electric power 光伏电缆

photovoltaic generation device 光伏发电设备

PHS (personal handphone system) 个人手持电话系统，双向无绳电话系统(日本)

PHY (physical) 物理，物理层

physical (PHY) 物理，物理层

physical channel 物理信道

physical circuit 物理线路

physical construction 物理构造

physical contact (PC) 物理接触

physical damage 物理损害

physical keying 物理锁键

physical layer convergence protocol (PLCP) 物理层会聚协议

physical layer media dependent (PLMD) 物理层相关媒体

physical length 物理长度

physical media dependent (PMD) 物理层相关媒体

physical medium dependent sublayer 物理介质相关子层

physical protection 实体防范

physical resolution 物理分辨率

Physical Security Interoperability

Alliance (PSIA) 物理安防互操作性联盟
physical separation 物理间隔[分离]
PI controller 比例积分控制器
picture 图像,模式
picture definition 图像清晰度
picture element 像素
picture jitter 图像晃动
picture reordering 图像重排序
picture resolution 图像分辨率
picture roll 图像滚动
picture slip 图像滑动
picture transfer protocol (PTP) 图片传输协议
picture-in-picture (PIP) 画中画
picture-sound desynchronization 声画失调
picture-sound lag 声画延迟
PID (packet identifier) 包标识符
PID (passive infrared detector) 被动红外探测器
PID (proportional integral derivative) 比例,积分,微分
pigtail 尾纤
pilot sub-carrier 导频子载波
PIM (protocol independent multicast) 独立于协议的多播
PiMF 对对铝箔屏蔽
PIM-SM (protocol independent multicast-sparse mode) 稀疏模式独立组播协议
pin 插针,引脚
PIN (personal identification number) 个人身份识别码
PIN (program item number) 节目片段码
pin tumbler lock 弹子锁
pinhole camera 针孔摄像机
pin-pair assignment 引脚分配
PIP (picture-in-picture) 画中画
pipe 管道
pipe bender 弯管机
pipe of leading to ground 引地管道
pipeline 管路,流水线
pipeline gas fire extinguishing system 管网气体灭火系统
pipeline laying 管路敷设
pipeline sampling suction smoke fire detector 管路采样吸气式感烟火灾探测器
piping assembly 配管
piping system 管道系统
PIR (passive infrared) 被动红外
piston wind 活塞风
pitch 音高,节距
pixel 像素,像元
pixel defect 像素缺陷
pixel density 像素密度

pixel pitch 像素中心距,像素间距
pixelation 像素化
PKC (public key certificate) 公钥证书
PL (permanent link) 永久链路
place 场所
place of origin 原产地
plain old telephone service (POTS) 老式电话服务
plan 平面图
planning and installation 规划和安装
plasma display panel (PDP) 等离子显示器,电浆显示板
plastic cable support system 塑料电缆桥架
plastic housing 塑料外壳
plastic insulated control cable 塑料绝缘控制电缆
plastic optical fiber (POF) 全塑[塑料]光纤
plastic spray 喷塑
plastic sub-pipe 塑料子管
plastic tape 塑胶带
platform 站台
play range 播放区域
playout system 播出系统
PLC (power line carrier) 电力线载波通信
PLC (programmable logic control) 可编程逻辑控制
PLC (programmable logic controller) 可编程逻辑控制器
PLC-BUS 电力线通信总线技术
PLCP (physical layer convergence protocol) 物理层会聚协议
plesiochronous digital hierarchy (PDH) 准同步数字系列
PLMD (physical layer media dependent) 物理层相关媒体
PLMN (public land mobile network) 公共陆地移动网
PLMS (parking lot management system) 停车库(场)管理系统
PLMT (power load management terminal) 电力负荷管理终端
PLP (packet layer protocol) 分组层协议
plug 插头
plug insertion life 插拔寿命
plug-and-play (PnP) 即插即用
pluggable authentication modules (PAM) 可插入认证模块
pluggable UPS-type A A型插接式UPS
pluggable UPS-type B B型插接式UPS
plug-in module 插件
PMC (PCI mezzanine card) PCI夹层卡

P

PMD (physical media dependent) 物理层相关媒体

PMD (polarization mode dispersion) 偏振[极性]模色散

PMO (project management office) 项目管理办公室

PMP (project management plan) 项目管理计划

PMS (parking management system) 停车管理系统

PMS (patient monitoring system) 病人监护系统

PMT (program map table) 节目映射表

PN (personal number) 个人号码

PnP (plug-and-play) 即插即用

POE (power over Ethernet) 以太网供电，有源以太网

POF (plastic optical fiber) 全塑[塑料]光纤

point coordination function (PCF) 点协调功能

point flame detector 点型火焰探测器

point guarding 点警戒

point of access to a pathway 路径接入点

point-to-multipoint (P2MP) 点到多点

point-to-multipoint multicast (PTM-M) 点对多点组播业务

point-to-multipoint service center (PTM-SC) 点到多点时局服务中心

point-to-point (P2P) 点到点

point-to-point protocol (PPP) 点到点协议

point-to-point protocol over Ethernet (PPPoE) 基于以太网的点对点协议

point-to-point tunneling protocol (PPTP) 点对点隧道协议

point-to-point data link 点对点数据链路

point-type carbon monoxide fire detector 点型一氧化碳火灾探测器

point-type ion smoke detector 点型离子感烟(火灾)探测器

point-type photoelectric smoke detector 点型光电感烟(火灾)探测器

point-type sampling aspiration smoke detector 点型采样吸气式感烟(火灾)探测器

point-type smoke fire detector 点型感烟火灾探测器

point-type temperature fire detector 点型感温火灾探测器

POL (power over LAN) 基于局域网的供电系统

P

polarisation mode dispersion (PMD) 偏振模色散
polarity 极性
polarization diversity receiver (PDR) 偏振[极化]分集接收机
polarization mode dispersion (PMD) 偏振[极化]模色散
polarized antenna 极化天线
polarizer 偏振片
policy certification authorities (PCA) 认证管理机构
pollution 污染
polycarbonate (PC) 聚碳酸酯
polyester foil 聚酯薄膜，聚酯带
polyethylene (PE) 聚乙烯
polypropylene insulated telephone cord 聚丙烯绝缘电话软线
polyurethane (PUR) 聚氨酯
polyvinyl chloride (PVC) 聚氯乙烯
polyvinyl chloride insulated and sheathed control cable 聚氯乙烯绝缘和护套控制电缆
polyvinyl chloride insulated cable 聚氯乙烯绝缘电缆
polyvinyl chloride insulated telephone cord 聚氯乙烯绝缘电话软线
PON (passive optical network) 无源光网络
POP (post office protocol) 邮局协议
porous media 多孔介质
portable document format (PDF) 便携式文档格式
portable operating system interface of UNIX (POSIX) 可移植操作系统接口
POS (passive optical splitter) 无源光纤分光器
positive 正面
POSIX (portable operating system interface of UNIX) 可移植操作系统接口
post office protocol (POP) 邮局协议
post telephone & telegraph 邮政电话和电报
posterization 多色调分色法
pot head 电缆终端套管
potential 电位
potential difference 电位差
potential transformer (PT) 电压互感器，比压器
POTS (plain old telephone service) 老式电话服务
powder coating 粉末涂料
power amplifier 功率放大器
power cable 电力[动力]电缆
power capacity 功率容量
power cord of cabinet 机柜电源线

power cord trough 电源线槽

power distribution monitoring and control 配电监控

power distribution panel (PDP) 配电盘

power distribution room 变配电室

power factor 功率因子

power factor correction (PFC) 功率因数校正

power failure 电源故障

power indicator 电源指示灯

power line 电力线

power line carrier (PLC) 电力线载波通信

power load management terminal (PLMT) 电力负荷管理终端

power meter 电度表,功率计

power over Ethernet (POE) 以太网供电,有源以太网

power over LAN (POL) 基于局域网的供电系统

power response 功率响应

power sum ACR (PS ACR) 功率和 ACR

power sum alien (exogenous) far-end crosstalk loss (PS AFEXT) 远端外部串扰损耗功率和

power sum alien near-end crosstalk (loss) (PS ANEXT) 外部近端串音功率和,近端外部串扰损耗功率和

power sum attenuation to alien (exogenous) crosstalk ratio at the far-end (PS AACR-F) 远端衰减与外部串扰比功率和

power sum attenuation to alien (exogenous) crosstalk ratio at the near-end (PS AACR-N) 近端衰减与外部串扰比功率和

power sum attenuation to crosstalk ratio at the far-end (PS ACR-F) 远端衰减与串扰比功率和

power sum attenuation to crosstalk ratio at the near-end (PS ACR-N) 近端衰减与串扰比功率和

power sum ELFEXT attenuation (loss) (PS ELFEXT) 功率和 ELFEXT 衰减(损耗)

power sum equal level far-end crosstalk ratio (PS ELFEXT) 等电平远端串扰衰减功率和

power sum far end crosstalk (loss) (PS FEXT) 远端串音功率和

power sum FEXT attenuation (loss) (PS FEXT) 功率和 FEXT 衰减(损耗)

power sum NEXT attenuation (loss) (PS NEXT) 近端串音功率和

power supply 供电器,供电电源

power supply mode 供电方式

power supply outlet 电源插座
power supply system (PSS) 供电系统
power usage effectiveness (PUE) 电能利用效率
power wire 电源线
PP (patch panel) 配线架,跳线架,转接面板
PP (protection profile) 保护轮廓
PPC (pay per channel) 按频道付费
PPCH (packet paging channel) 分组寻呼信道
PPD (partial packet discard) 部分分组丢弃
PPM (pulse phase modulation) 脉冲位置调制
PPP (point-to-point protocol) 点到点协议
PPP (point-to-point protocol) 点到点协议
PPPoE (point-to-point protocol over Ethernet) 基于以太网的点对点协议
PPSN (public packet-switching network) 公共分组交换网络
PPTP (point-to-point tunneling protocol) 点对点隧道协议
PPV (pay per view) 按次付费
PR (packet retransmission) 分组转发
PRACH (packet random access channel) 分组随机接入信道
PRC (primary reference clock) 基准时钟
pre-assemble 预端接
pre-assembled fibre optic cable 预端接光缆
pre-assembly of breakout cable 预端接分支光缆
pre-assembly of loose-tube cable 预端接松套管光缆
precedence level 优先级
perception multimedia 感觉多媒体
precision air distribution 精确送风
precision time protocol (PTP) 精确时间协议
prediction 预测
prediction block 预测块
prediction compensation 预测补偿
prediction partition type 预测划分方式
prediction process 预测过程
prediction unit 预测单元
prediction value 预测值
predictive failure analysis (PFA) 故障预警分析
preferred source 首选电源

preformed hole 预留孔
preheat 预热
premises distribution system (PDS) 综合布线系统
premises owner 业主
premium rate (PRM) 附加费率
presentation channel 播放声道
presentation multimedia 表示媒体
presentation time stamp (PTS) 呈现时间戳
presentation unit (PU) 呈现单元
preset position 预置位
pressure gage 压力表
pressurization fan 增压风机
pre-terminated copper trunking cable 预端接集束铜缆
pre-terminated drop fiber cable 预制成端型引入光缆
pretty good privacy (PGP) 优良密保(协议)
prevention of burglary 防盗
prevention of fire 防火
preview tours 预览轮切
PRG (pseudo-random generator) 伪随机发生器
PRI (primary rate interface) 基群速率接口
primary circuit 主电路
primary colour 基色
primary colour picture 基色图像
primary colour signal 基色信号
primary distribution 一次分配
primary distribution space 主配线空间
primary entrance room 主进线间
primary power 主电源
primary rate interface (PRI) 基群速率接口
primary reference clock (PRC) 基准时钟
prime lens 定焦
printer 打印机
priority of service 服务优先级
privacy lock 保密锁
private automatic branch exchange (PABX) 专用自动交换分机
private branch exchange (PBX) 程控用户交换机
private data network (PDN) 专用数据网
private network 专[私]网
PRM (premium rate) 附加费率
procedure 规程
process automation (PA) 过程自动化
process CPG CPG进程
process quality 工艺质量
processing amplifier 处理放大器
processing element memory (PEM) 处理元件存储器,局部存储器

P

processor 处理器
produce 生产
product catalogue 产品目录
product inspection record 产品检验记录
product life 产品寿命
product life testing 产品寿命测试
product manufacturer 产品制造商
product protection 产品保护
product sample 产品样本
production and broadcast system based on computer and network 基于计算机与网络的制播系统
profile 配置文件
profile chart 剖面图
program 节目
program associated data (PAD) 节目相关数据
program association table (PAT) 节目关联表
program bus 程序总线
program clock reference (PCR) 节目时钟基准
program content evaluation 节目内容评估
program dynamic range (PDR) 节目(音频)动态范围
program element 节目元素
program information 节目信息
program item 节目项
program item number (PIN) 节目片段码
program map table (PMT) 节目映射表
program memory 程序存储器
program service 节目业务
program specific information (PSI) 节目特定信息
program stream (PS) 节目流
program transmission equipment 节目传输设备
program type (PTY) 节目类型
program type name (PTYN) 节目类型名
programmable computer controller (PCC) 可编程计算机控制器
programmable logic control (PLC) 可编程逻辑控制
programmable logic controller (PLC) 可编程逻辑控制器
progressive scanning 逐行扫描
project management 项目管理
project management office (PMO) 项目管理办公室
project management plan (PMP) 项目管理计划
project manager 项目经理
project organization 项目组织
project progress plan 工程进度计划

project scope 工程范围
project supervision 项目监理机构
projection screen 投影屏幕
projector 投影机
proof test for fire-extinguishing system 消防系统效用试验
propaganda fire vehicle 宣传消防车
propagation delay 传播时延
propagation delay skew 延时差
propane detector 丙烷探测器
proportional integral derivative (PID) 比例、积分、微分
protected 受保护的
protected area 防护区
protected stairway 疏散楼梯
protection apparatus 保护装置
protection area 防护区
protection factor of protective device 保护装置的保护因数
protection level 保障等级
protection object 防护对象
protection profile (PP) 保护轮廓
protection rating 防护等级
protective conductor 保护导体
protective conductor current 保护导体电流
protective cover 防护罩
protective coverings for electric cable 电缆外护层
protective earth and neutral conductor (PEN conductor) PEN导体
protective earthing 保护性接地
protective earthing conductor (PE) 保护性接地导体
protective housing 保护壳体
protective interlock 保安联锁装置
protective jacket 保护罩
protective layer 保护层
protective system 安全系统
protective tape 保护带
protective window 防护窗
protector 保安器
protector block 保安器组件
protein foam concentrate 蛋白泡沫液，蛋白泡沫浓缩物
protocol analyzer 协议分析仪
protocol data network 协议数据网
protocol independent multicast (PIM) 独立于协议的多播
protocol independent multicast-sparse mode (PIM-SM) 稀疏模式独立组播协议
proximity card 感应[接近]卡
PRS (pseudo-random sequence) 伪随机序列
PS (packet switching) 分组[分包]交换
PS (program stream) 节目流
PS AACR-F (power sum attenuation to alien/exogenous crosstalk ratio

at the far-end) 远端衰减与外部串扰比功率和

PS AACR-Favg (average power sum attenuation to alien/exogenous crosstalk ratio far-end) 外部远端串扰损耗功率和平均值

PS AACR-N (power sum attenuation to alien/exogenous crosstalk ratio at the near-end) 近端衰减与外部串扰比功率和

PS ACR (power sum ACR) 功率和 ACR

PS ACR-F (power sum attenuation to crosstalk ratio at the far-end) 远端衰减与串扰比功率和

PS ACR-N (power sum attenuation to crosstalk ratio at the near-end) 近端衰减与串扰比功率和

PS AFEXT (power sum alien/exogenous far-end crosstalk loss) 远端外部串扰损耗功率和

PS ANEXT (power sum alien near-end crosstalk loss) 近端外部串扰损耗功率和

PS ANEXTavg (average power sum alien near-end crosstalk loss) 外部近端串音功率和平均值

PS ELFEXT (power sum equal level far-end crosstalk ratio) 等电平远端串扰衰减功率和

PS FEXT (power sum far end crosstalk loss) 远端串扰损耗功率和

PS NEXT (power sum NEXT attenuation loss) 近端串扰损耗功率和

PSDN (packet switched data network) 公共分组数据网络

PSE (packet switching exchange) 分组交换机

pseudo-random 伪随机

pseudo-random code 伪随机码

pseudo-random generator (PRG) 伪随机发生器

pseudo-random sequence (PRS) 伪随机序列

PSI (program specific information) 节目特定信息

PSIA (Physical Security Interoperability Alliance) 物理安防互操作性联盟

PSK (phase shift keying) 相移键控

PSNR (peak signal to noise ratio) 峰值信噪比

PSPDN (packet switched public data network) 分组交换公用数据网

PSS (power supply system) 供电系统

PSTN (public switched telephone network) 公共交换电话网络

PT (packet terminal) 分组终端
PT (payload type) 有效载荷类型
PT (potential transformer) 电压互感器,比压器
PTCCH (packet timing advanced control channel) 分组定时提前控制信道
PTM (performance test model) 性能测试模型
PTM-M (point-to-multipoint multicast) 点对多点组播业务
PTM-SC (point-to-multipoint service center) 点到多点时局服务中心
P-TMSI (packet TMSI) 分组临时移动用户识别码,分组 TMSI
PTN (personal telecommunication number) 个人通信号码
PTO (public telecommunication operator) 公众电信运营商
PTP (picture transfer protocol) 图片传输协议
PTP (precision time protocol) 精确时间协议
PTS (presentation time stamp) 呈现时间戳
PTS (public telecommunications service) 公用电信服务
PTY (program type) 节目类型
PTYN (program type name) 节目类型名
PTZ (pan/tilt/zoom) 云台全方位控制
PTZ cruise 云台巡航
PU (peripheral unit) 外围单元[设备]
PU (presentation unit) 呈现单元
public address (PA) 公共广播
public address equipment 公共广播设备
public address system (PAS) 公共广播系统
public building 公共建筑
public data network (PDN) 公共数据网络
public data transmission service 公众数据传输业务
public emergency 突发公共事件
public key certificate (PKC) 公钥证书
public land mobile network (PLMN) 公共陆地移动网
public management center 公共管理中心
public network (PDN) 公共[公用]网络
public network provider 公共网络提供商
public packet-switching network (PPSN) 公共分组交换网
public part 公共部位

public place 公共场所

public security system 公共安全系统

public switched telephone network (PSTN) 公共交换电话网络

public telecommunication operator (PTO) 公众电信运营商

public telecommunications service (PTS) 公用电信服务

published user data 公开用户数据

PUE (power usage effectiveness) 电能利用效率

pull-push permanent dust-proof cover 推拉式永久防尘盖

pulse amplitude modulation (PAM) 脉(冲)幅(度)调制

pulse code modulation (PCM) 脉(冲编)码调制

pulse counter 脉冲计数器

pulse phase modulation (PPM) 脉冲位置调制

pulse width modulation (PWM) 脉宽调制

pump laser 泵浦激光器

PUR (polyurethane) 聚氨酯

push VOD 推送式视频点播

PVC (permanent virtual circuit) 永久虚拟线路

PVC (permanent virtual connection) 永久虚连接

PVC (polyvinyl chloride) 聚氯乙烯

PVC insulated ribbon cable 聚氯乙烯绝缘带状电缆

PVC insulation 聚氯乙烯绝缘

PVC/ST4 聚氯乙烯绝缘电缆、固定敷设用电缆护套

PWM (pulse width modulation) 脉宽调制

PXC (photonic cross connect) 全光交叉连接

Q

Q3 (GSM Q3 protocol) GSM 的 Q3 协议

QA (quality assurance) 质量保证

QAF (quality assessment form) 质量评定表

QAM (quadrature amplitude modulation) 正交调幅，正交振幅调制器

QAM signal (quadrature amplitude modulation signal) QAM[正交调幅]信号

QC (quality control) 质量控制

QCIF (quarter common intermediate format) 四分之一通用媒介格式

QCM (quality control manual) 质量管理手册

QCP (quickly connected plug) 快接插头

Q-Factor (quality factor) 品质因数

QMP (quality management plan) 质量管理计划

QoS (quality-of-service) 服务质量

QRM (quad relay module) 四继电器模块

QSFP (quad small form-factor pluggable) 四通道 SFP 接口

quad 四芯电缆，四线组

QUAD (quadriphonic) 四声道的，四声道立体声的

quad relay module (QRM) 四继电器模块

quad small form-factor pluggable (QSFP) 四通道 SFP 接口

quadrant (QD) 象限

quadrature amplitude modulation (QAM) 正交调幅，正交振幅调制器

quadrature amplitude modulation signal (QAM signal) 正交调幅调制信号

quadrature modulation 正交调制

quadriphonic (QUAD) 四声道的，四声道立体声的

quality assessment form (QAF) 质

量评定表

quality assessment table 质量评定表

quality assurance (QA) 质量保证

quality assurance certificate 质量保证书

quality assurance period 质量保证期

quality assurance schematic 质量保证流程图

quality certificate 产品合格证

quality certification 质量认证

quality control (QC) 质量控制

quality control manual (QCM) 质量管理手册

quality factor (Q-Factor) 品质因数

quality inspection measure 质量检测措施

quality management plan (QMP) 质量管理计划

quality-of-service (QoS) 服务质量

quality plan 质量计划

quality report in guarantee period 质保期报告

quality system 质量体系

quality warranty 质量保证书

QUANT (quantitative) 定量的

quantitative (QUANT) 定量的

quantization coefficient 量化系数

quantization parameter 量化参数

quantize (QUANT) 量化

quarter common intermediate format (QCIF) 四分之一通用媒介格式

quick connection technology 快速连接技术

quick identification 快速识别

quick response code (QR code) 二维码

quickly connected plug (QCP) 快接插头

R

raceway 电缆管道
RACH (random access channel) 随机接入信道
rack fixture set 机架装配套件
rack kit 机柜套件
rack metalwork 机架金属件
rack space 储存架区
rack unit 机架单元
radial shrinkage ratio 径向收缩率
radiant energy-sensing fire detector 辐射能感应火灾探测器
radiant heating 辐射加热
radiant resistance furnace 辐射电阻炉
radiation 辐射
radiation detection 辐射探测器
radiation fire detector 辐射火灾探测器
radiator cooler 辅助冷却装置
radiator unit 散热部件
radio access network (RAN) 无线接入网络
radio alarming 无线报警
radio antenna 无线电天线
radio base station (RBS) 无线基站
radio beacon 无线电信标机
radio broadcasting center system 广播中心系统
radio cache 无线电贮藏处
radio channel 无线电信道
radio channel group (RCG) 无线信道群
radio circuit 无线电电路
radio communication 无线电通信
radio communication system 无线电通信系统
radio configuration 无线配置
radio data system (RDS) 无线数据系统
radio dispatch system 无线电调度系统
radio fix 无线电定位
radio frequency (RF) 射频
radio frequency cable 射频电缆
radio frequency coaxial connector 射频同轴连接器

radio frequency identification (RFID) 射频识别
radio frequency signal-noise ratio (RFSNR) 射频信噪比
radio interface layer 3 (RIL 3) 无线接口层 3
radio interface protocol 无线接口协议
radio interference 无线电干扰
radio jamming 无线电干扰
radio link 无线电链路[线路]
radio link control (RLC) 无线链路控制
radio link management (RLM) 无线链路管理
radio link protocol (RLP) 无线链路协议
radio mast 无线电杆,天线杆
radio network controller under the UMTS system UMTS 系统下的无线网络控制器
radio OSI protocol 无线 OSI 协议
radio pager 无线电寻呼机
radio paging 无线电寻呼
radio paging system 无线寻呼系统
radio port controller (RPC) 无线端口控制器
radio relay 无线电中继站
radio resource (RR) 无线资源
radio set 无线电台
radio sub-system criteria 无线电子系统标准,无线分系统准则
radio test equipment (board) (RTE) 无线测试设备的射频单元
radio traffic 无线电通信
radio transmitting tower 无线电发射塔
radio warning 无线电报警
RADIUS (remote authentication dial in user service) 远程用户拨号认证系统
RAID (redundant array of independent disk) 磁碟阵列,独立冗余磁碟阵列
RAID controller RAID 控制器
RAID0 RAID 级别 0
RAID1 RAID 级别 1
RAID2 RAID 级别 2
RAID3 RAID 级别 3
RAID4 RAID 级别 4
RAID5 RAID 级别 5
RAID6 RAID 级别 6
RAID10 RAID 级别 10
RAID50 RAID 级别 50
rail adapter 导轨适配器
rail-mounted 轨道安装
railway tunnel 铁路隧道
raised floor 高架[活动]地板
rake receiver 瑞克接收机,分离多

径接收机

RAM (random access memory) 随机存取存储器

ramp 坡道

ramped aisle 斜坡通道

RAN (radio access network) 无线接入网

random access 随机接入,随机访问

random access channel (RACH) 随机接入信道

random access memory (RAM) 随机存取存储器

Rangmuir equation 郎缪尔公式

Rankine cycle 兰金循环

Ranque-Hilsch effect 兰克-赫尔胥效应

rapid ring protection protocol (RRPP) 快速环网保护协议

rapid spanning tree protocol (RSTP) 快速生成树协议

rapper 振动器

RAS (remote access service) 远程访问服务

raster scan 光栅扫描

rate 码率,变化率

rate of combustion 燃烧效率

rate of cooling 冷却速度

rate of expansion 膨胀率

rate of filtration 过滤(速)率

rate of flow 流速

rate of propagation 传播速度

rate of radiation 辐射强度

rate of revolution 转速

rate of vaporization 汽化率

rate updating index (RID) 费率修改索引

rated cooling capacity 额定制冷量

rated current 额定电流

rated energy efficiency grade 额定能效等级

rated fire door 额定防火门

rated frequency 额定频率

rated frequency range 额定频率范围

rated load 额定负荷

rated of air circulation 空气循环额定值,换气次数,空气循环率

rated of decay 衰减率

rated output active power 额定输出有功功率

rated output apparent power 额定输出视在功率

rated partition 耐火等级隔墙

rated power 额定功率

rated transmission voltage 额定传输电压

rated value 额定值

rated voltage 额定电压

rated voltage range 额定电压范围

R

rated wall 耐火等级墙

rate-of-rise and fixed temperature detector 差定温探测器

rate-of-rise detector 差温探测器

rating 额定值

ratio controller 比例调节器

ratio meter 比率计

ratio of expansion 膨胀比

ratio of run-off 径流系数,径流比

ratio of slope 坡度

ratio of specific heat 比热比,热容比

ratio of windows to wall 窗墙比

raw data 原始数据

raw material 原材料

ray radiation 光辐射

Rayleigh channel 瑞利信道

Rayleigh fading 瑞利衰落

RBAC (role-based access control) 基于角色的访问控制,基于任务的访问控制

RBS (radio base station) 无线基站

RC (remote control) 远程控制,遥控

RCE (remote control equipment) 远端控制设备

RCG (radio channel group) 无线信道群

RCM 柔性化制造单元自动化控制调度系统

RCWS (recirculating cooling water system) 循环冷却水系统

RDB (relational database) 关系数据库

RDBMS (relational database management system) 关系数据库管理系统

RDI (remote defect indication) 远端缺陷指示

RDS (radio data system) 无线数据系统

reach in refrigerator 大型冷柜

reaction of replacement 置换反应

reactive power 无功功率

reactor 反应堆

reader 读卡器

ready-to-install fibre optic multiple cable 预端接集束光缆

real time (RT) 实时

real time clock (RTC) 实时时钟

real time differential (RTD) 实时动态码相位差分技术

real time streaming protocol (RTSP) 实时流媒体协议

real-time alarm 实时告警

real-time intelligent patch cord management system 实时智能跳线管理系统

real-time transport control protocol (RTCP) 实时传输控制协议

real-time transport protocol (RTP) 实时传输协议
rear screen projection 背[后]投影
Reaumur 列氏温度计
rebroadcasting 转播
re-cabling 再布缆
receive 接收
receive diversity 接收分集
receive power dynamic range 接收功率动态范围
received signal strength indication (RSSI) 接收信号强度指示
receiver (RX) 接收机
receiver C/(N+1) 接收机C/(N+1)
receiver selectivity 接收机选择性
receiver tuning range 接收机调谐范围
receiving antenna 接收天线
receiving basin 蓄水池
receiving end 接收端
receptacle outlet box 接线盒
reception area 接待处
reception of heat 吸热
recessed radiator 暗装散热器
recessed sprinkler system 半隐蔽型喷水灭火系统
recharge well 回灌井
reciprocal compressor 往复式压缩机
reciprocal feed pump 往复给水泵
reciprocal grate 往复炉排
reciprocal proportion 反比例
reciprocating compressor 往复式压缩机
reciprocating refrigerator 往复式制冷机
recirculated air 再循环空气
recirculating cooling water system (RCWS) 循环冷却水系统
reclaimed water 再生水
reclamation of condensate water 蒸汽冷凝水回收
reclocking 时基重建
recognition process 认可过程
reconnection 多次端接
reconstruct 改建
reconstructed sample 重建样本
record 记录
record and playback 记录和回放
recorded broadcast 录播
recording and production system 录制系统
recording apparatus 记录仪器
recording barometer 自记气压计
recording facility 记录装置
recording liquid level gauge 自动液面计
recording pressure gauge 自记压力计

R

recording water-gauge 自记水位计
recovery time 恢复时间
rectification 整改，校正
recyclable material 可再循环材料
red green blue (RGB) RGB 色彩模式
red green blue horizontal vertical (RGBHV) RGB 与水平、垂直信号
reduced instruction set computing (RISC) 精简指令集计算
redundancy 冗余，冗余度，冗余信息
redundant array of independent disk (RAID) 磁碟阵列，独立冗余磁碟[磁盘]阵列
redundant power system (RPS) 冗余电源系统
redundant system 冗余系统
Reed-Solomon code 里德-所罗门码
reed switch 干簧管
reel cleaner 卷轴清洁器
reference building 参照建筑
reference field 参考场
reference frame 参考帧
reference frequency 参考[参比]频率
reference index 参考索引
reference magnetic level 参考磁平
reference picture 参考图像
reference picture buffer (RPB) 参考图像缓冲区
reference picture list (RPL) 参考图像队列
reference service 参考业务
reference target 参考目标
reference testing 参考测试
reference white level code value 参考白电平编码值
referenced cabling design document 参考的布缆设计文档
reflection 反射
reflective coefficient of solar radiation 太阳能辐射反射系数
reflective detector 反射式探测器
reflective plate 反射板
reflux valve (RV) 止回阀
refractive 折射
refractive index 折射系数
refractory copper wire and cable 耐火铜芯电线电缆
refractory fiber material 耐火纤维
refresh frequency 刷新频率
refresh rate 刷新率
refrigerator room 冷冻机房
refuge floor 避难层
region of interest (ROI) 感兴趣区域
regional alarm system 区域报警

系统

registered trademark 注册商标

registration 登记

regular pulse excited long-term prediction (RPE-LTP) 规则脉冲激励长期预测

reinforcing rod 加强筋

related layer 相关层

relational database (RDB) 关系数据库

relational database management system (RDBMS) 关系数据库管理系统

relative humidity (RH) 相对湿度

relative permeability 相对磁导率

relay 继电器

reliability check 可靠性检查

relief damper 泄压风门

remote access 远程访问

remote access service (RAS) 远程访问服务

remote access software 远程访问软件

remote alarm processing center 远程联网报警中心

remote authentication dial in user service (RADIUS) 远程用户拨号认证系统

remote control (RC) 远程控制，遥控

remote control and monitor 远程监控

remote control equipment (RCE) 远端控制设备

remote control microphone 遥控传声器

remote control of connecting technology 远程监控连接技术

remote defect indication (RDI) 远端缺陷指示

remote job entry (RJE) 远程作业输入

remote monitoring 远程监控

remote monitoring MIB 基于MIB的远程监控

remote monitoring system (RMS) 远程监控系统

remote node (RN) 远程节点

remote point 远端点

remote power feeding (RPF) 远程供电

remote powered device (RPD) 远程供电设备

remote powering 远程供电

remote procedure call (RPC) 远程过程调用

remote site 远程站点

remote socket 遥控插座

remote subscriber unit/line element 远端接入局(站)，远端用户单

R

元(或线路终端单元)

remote terminal unit (RTU) 远程终端单元,远程终端控制系统

removable connection 活动连接

renewable energy 可再生能源

renewable energy source (RES) 可再生能源

renewal of the blower fan coil 风机盘管加新风

repeated frame 重复帧

repeater (RP) 中继器,重发器,增音器

report server 报表服务[伺服]器

representational state transfer (REST) 代表性状态传输,表述性状态传递

request for comment (RFC) 请求注释[评论]

request to exit (RTE) 请求退出

RES (renewable energy source) 可再生能源

RES (reserved) 保留

research of requirement 需求调研

reserved (RES) 保留

reserved bit 保留位

reset circuit (RSC) 复位电路

reset request (RSR) 复位请求

residential 住宅

residential area 居住区

residential area information system 居住区信息系统

residential block mobile telecommunication equipment room 小区移动电信设备室[机房]

residential building 住宅,居住建筑

residual 残差

residual current operated protective device 剩余电流保护装置

residual current-type electrical fire monitoring detector 剩余电流式电气火灾监控探测器

resistance temperature detector (RTD) 电阻温度探测器,电阻温度传感器

resistance to earth 对地电阻

resolution 分辨率

resolution enhancement technology (RET) 分辨率增强技术

resource allocation 资源分配

resource system 资源系统

responding channel 应答通道

response time index (RTI) 响应时间指数

response/deviation 响应与偏差

REST (representational state transfer) 代表性状态传输,表述性状态传递

restored energy time 能量恢复

时间

restricted area 禁区

restricted mode launch (RML) 限模注入

restricted mode launch bandwidth (RML bandwidth) 限模注入带宽

restricted mode launch measurement (RML measurement) 限模注入测试

restriction of hazardous substances (RoHS) (电气、电子设备中)限制使用某些有害物质指令

RET (resolution enhancement technology) 分辨率增强技术

return air 回风口

return air flame plate 回风百叶

return loss (RL) 回波损耗(回损)

return-to-zero (RZ) 归零制

return-to-zero binary code 归零二进制码

re-useable material 可再利用材料

re-use factor 重用系数

re-useable splice 可反复使用的端接点

reverberation 混响

reverberation time 混响时间

reverse link 反向链接

RF (radio frequency) 射频

RF amplification RF[射频]放大

RF amplifier RF[射频]放大器

RF conversion RF[射频]转换

RF cutoff RF[射频]关断

RF distribution RF[射频]分配

RF input/output return loss RF[射频]输入输出反射损耗

RF output on/off ratio RF[射频]输出关断度

RF over lay RF 混合,射频叠加

RF scramble 射频加扰

RF wanted-to-interfering signal ratio 射频信扰比

RF64 RIFF 64 位版本

RFC (request for comment) 请求注释[评论]

RFID (radio frequency identification) 射频识别

RFSNR (radio frequency signal-noise ratio) 射频信噪比

RGB (red green blue) RGB 色彩模式

RGBHV (red green blue horizontal vertical) RGB 与水平、垂直信号

RH (relative humidity) 相对湿度

ribbon cable 带状电缆

RID (rate updating index) 费率修改索引

RIL 3 (radio interface layer 3) 无线接口层 3

ring 环形

ring network 环形网络

RIP (routing information protocol) 路由信息协议

ripcord 撕裂[剥离]绳

RIPng (routing information protocol, next generation) 下一代路由协议

RISC (reduced instruction set computing) 精简指令集计算

riser backbone subsystem 垂直干线子系统

riser cable 垂直布线电缆

risk analysis 风险分析

risk assessment 风险评估

rivet 铆钉

RJ45 RJ45 型连接器件

RJ45 module RJ45 型模块

RJ45 plug RJ45 插头

RJ45 socket unshielded 非屏蔽 RJ45 插座

RJ45-RJ45 unshielded RJ45 jumper wire RJ45-RJ45 非屏蔽 RJ45 跳线

RJE (remote job entry) 远程作业输入

RL (return loss) 回波损耗(回损)

RLC (radio link control) 无线链路控制

RLM (radio link management) 无线链路管理

RLP (radio link protocol) 无线链路协议

RML (restricted mode launch) 限模注入

RML bandwidth (restricted mode launch bandwidth) 限模注入带宽

RML measurement (restricted mode launch measurement) 限模注入测试

RMS (remote monitoring system) 远程监控系统

RMS (root mean square) 均方根值

RMS voltage variation 方均根电压变化

RN (remote node) 远程节点

roaming 漫游

roaming service 漫游服务

rodent protection 防鼠啮

rodent-resistant net 防鼠网

RoHS (restriction of hazardous substances) (电气、电子设备中)限制使用某些有害物质指令

ROI (region of interest) 感兴趣区域

role-based access control (RBAC) 基于角色的访问控制,基于任务的访问控制

rollback 回退
roof 屋脊
roof resisting to external fire exposure 防火屋面
roof screen 闷顶隔板
room 室,房间,机房
room environment 机房环境
room infrastructure 机房基础设施
room number 房号
root mean square (RMS) 均方根值
route processor 路由处理器
route switch module (RSM) 路由交换模块
route/switch processor 路由交换处理器
router 路由器
routing 路由
routing information protocol (RIP) 路由信息协议
routing information protocol, next generation (RIPng) 下一代路由协议
routing scheme 路由图
routing table 路由选择表
RP (repeater) 中继器,重发器,增音器
RPB (reference picture buffer) 参考图像缓冲区
RPC (radio port controller) 无线端口控制器
RPC (remote procedure call) 远程过程调用
RPD (remote powered device) 远程供电设备
RPE-LTP (regular pulse excited long-term prediction) 规则脉冲激励长期预测
RPF (remote power feeding) 远程供电
RPL (reference picture list) 参考图像队列
RPS (redundant power system) 冗余电源系统
RR (radio resource) 无线资源
RRPP (rapid ring protection protocol) 快速环网保护协议
RS-232 RS-232标准接口(串行通信接口标准之一)
RS-422 RS-422协议(全称为“平衡电压数字接口电路的电气特性”)
RS-485 RS-485协议(工业总线网络之一)
RSC (reset circuit) 复位电路
RSM (route switch module) 路由交换模块
RSR (reset request) 复位请求
RSSI (received signal strength

R

indication) 接收信号强度指示

RSTP (rapid spanning tree protocol) 快速生成树协议

RT (real time) 实时

RTC (real time clock) 实时时钟

RTCP (real-time transport control protocol) 实时传输控制协议

RTCP (RTP control protocol) RTP控制协议

RTD (real time differential) 实时动态码相位差分(技术)

RTD (resistance temperature detector) 电阻温度探测器,电阻温度传感器

RTE (radio test equipment board) 无线测试设备的射频单元

RTE (request to exit) 请求退出

RTI (response time index) 响应时间指数

RTP (real-time transport protocol) 实时传输协议

RTP clock RTP时钟

RTP control protocol (RTCP) RTP控制协议

RTP session RTP会话

RTP stream RTP流

RTSP (real time streaming protocol) 实时流媒体协议

RTU (remote terminal unit) 远程终端单元,远程终端控制系统

rubber insulated cable 橡皮绝缘电缆

rubber insulated telephone cord 橡皮绝缘电话软线

rubber universal wheel 橡皮万向轮

run 游程;运行

rural service area 农村服务区

rust proof capability 防锈能力

RV (reflux valve) 止回阀

RVS 铜芯聚氯乙烯绝缘绞型连接用软电线

RVV 铜芯聚氯乙烯绝缘聚氯乙烯护套软电线

RVVP 铜芯聚氯乙烯绝缘屏蔽聚氯乙烯护套软电缆

RX (receiver) 接收机

RZ (return-to-zero) 归零制

S

S frame S帧

S/FTP 金属箔线对屏蔽与金属编织网总屏蔽对绞电缆

S/FTQ 金属编织网总屏蔽与金属箔线对屏蔽四芯对绞电缆

S/PDIF (SONY/PHILIPS digital interface format) SONY、PHILIPS数字接口格式

SA (security automation) 安保自动化

SA (supply air) 送风

SACCH (slow associated control channel) 慢速随路控制信道

safe allowable floor load 楼板安全允许荷载

safe area 安全区域

safe capacity 安全承载能力

safe circuit 安全电路

safe construction 安全施工

safe control 安全控制

safe current 安全电流

safe egress 安全出口

safe escape 安全逃脱

safe evacuation 安全疏散

safe illumination 安全照明

safe refuge 避难所

safe strategy 安全策略

safeguard 安全保卫

safeguard construction 安全结构

safeguarding structure 防护构筑物

safety 安全(性)

safety appliance 保安[安全]装置

safety approval plate 安全合格牌照

safety assembly 安全装置

safety block 安全保护部件

safety circuit 安全电路

safety class 安全级别

safety communications equipment 安全通信设备

safety control circuit (SCC) 安全控制电路

safety control mark 安全控制标志

safety curtain 防火幕

safety cut-off 安全切断

safety design 安全设计

safety device 安全装置

safety door 安全门
safety door latch 安全门锁
safety engineer 安全工程师
safety engineering（SE） 安全工程学
safety equipment 安全设备
safety exhaust 安全排气阀
safety exit indicator light 安全出口指示标志灯
safety factor 安全系数
safety feature 安全装置
safety fuse cut-out 安全熔断器
safety gear 安全装置
safety glass 安全玻璃
safety ground 安全接地
safety grounding 安全接地，保护接地
safety hatch 安全舱口
safety helmet 安全帽
safety inspector 安全检查员
safety island 安全岛
safety ladder 安全梯
safety light 安全指示灯
safety lighting 安全照明
safety lighting fitting 安全照明装置
safety limit switch（SLS） 保险总开关
safety load 安全荷载
safety load factor 安全荷载系数
safety lock 安全锁
safety mark 安全标志
safety marking 安全标记
safety observation station（SOS） 安全观察站
safety of structure 结构安全性
safety plug 安全塞
safety precaution 安全预防措施
safety protection 安全防护
safety provision 安全措施
safety range 安全范围
safety relief 安全减压
safety relief valve（SRV） 安全减压阀
safety ring 安全环
safety rope 安全绳
safety rule 安全规程[规范]
safety screen 安全屏
safety service life 安全使用年限
safety shut-off valve 安全截止阀
safety shut-off valve 自动断路阀
safety sign 安全标志
safety signal 安全信号
safety spacing 安全间距
safety specification 安全规程[规范]
safety switch 安全开关
safety system 安全系统
safety technique 安全技术
safety trip valve 自动断路阀

safety valve 安全阀
safety voltage 安全电压
safety window 安全窗
safety zone 安全区
sample 样本
sample format 格式范例
sample project 样板段工程,样板项目
sample rate 采样率
sample rate conversion (SRC) 采样频率转换
sample value 样本值
sampling hole 采样孔
sampling pipe network 采样管网
sampling rate 采样率
sampling survey 抽样调查
SAN (storage area network) 存储区域网络
sandwich construction 夹层结构
sandwich pressurization system 分层加压系统
sandwich wall 夹心墙
sanitary sewer 下水道
SAP (subscriber access point) 用户接入点
SAR (segmentation and re-assembly sub-layer) 分段和重装子层
SAR (segmentation and reassembly) 分段与重组
SAS (security alarm system) 防盗报警系统
SAS (security automation system) 安保自动化系统
SAS (statistical analysis system) 统计分析系统
SAS (subscriber authorization system) 用户[订户]授权系统
sash (窗)框
SAT (system acceptance test) 系统验收测试
SATA (serial advanced technology attachment) 串行高级技术附件
SATCOM (satellite communication) 卫星通信
satellite antenna 卫星天线
satellite communication (SATCOM) 卫星通信
satellite news gathering (SNG) 卫星新闻采集
satellite phone 卫星电话
satellite television 卫星电视
saturated vapor pressure (SVP) 饱和蒸汽压
saturation 色饱和度
SBC (standard building code) 标准建筑规范
SBS (stimulated Brillouin scattering) 受激布里渊散射
SC (subscriber connector) 用户连接器

SC connector SC 型(光纤)连接器
SC duplex adaptor SC 双工适配器
scaffold 脚手架
scalable video coding (SVC) 可伸缩视频编码
scatter diagram 散布图
scatter gram 散布图
scatter plot 散布图
SCC (safety control circuit) 安全控制电路
SCC (supervisory computer control) 监督计算机控制
SCCS (supervisory computer control system) 计算机监控系统
SC-D (duplex SC connector) 双工 SC 连接器
SCD connector 双芯 SC(光纤)连接器
scene picture 场景图像
scene reference picture buffer 场景参考图像缓冲区
scenic lift 观景电梯
schedule 运行图,日程表
schematic 简图
schematic circuit 图式电路
schematic diagram 示意图
school 学校
SCI (short circuit isolator) 短路隔离器
scissors stair 剪式楼梯
scope of work 工作范围
SCP (service and content protection) 业务和内容保护
SCP (service concentration point) 服务集合[汇集]点
SCP (short circuit protection) 短路保护
SCR (silicon controlled rectifier) 可控硅整流器
scrambled television 加扰电视
scrambler 加扰器
scrambling 加扰
scrambling algorithm 加扰算法
screen continuity 屏蔽连续性
screen door 纱门
screen gain 屏幕增益
screened balanced cable 屏蔽对绞电缆
screened cable 屏蔽电缆
screened channel 屏蔽信道
screened connector 屏蔽连接器
screened pair 线对屏蔽
screening room 放映室
screw erection hole 螺丝安装孔
screw hole 螺丝孔
screw stair 螺旋楼梯
screw terminal 螺丝端子
screwed cable gland 螺纹线缆接头
scroll bar 滚动条
scrolling effect 滚动效果

SCS (**structure cabling system**) 结构化布线系统

ScTP (丝网总屏蔽/铝箔线对)屏蔽双绞线

scupper 排水洞[口]

scuttle hatch 屋顶楼板开口

SD (**service distributor**) 服务配线架

SD (**shading coefficient of window**) 外窗遮阳系数

SD extended capacity (**SDXC**) 容量扩大化的安全存储卡

SDCCH (**standalone dedicated control channel**) 独立(专用)控制信道

SDH (**synchronous digital hierarchy**) 同步数字分层结构,同步数字体系

SDHC (**secure digital high capacity**) SDHC安全数字大容量卡

SDI (**serial digital interface**) 串行数字接口

SDK (**software development kit**) 软件开发工具包

SDL (**specification description language**) 规范描述语言

SDMA (**space division multiple access**) 空分多址

SDP (**session description protocol**) 会话描述协议

SDSL (**single-line digital subscriber line**) 单线(路)数字用户线路

SDT (**service description table**) 业务描述表

SDTV (**standard definition television**) 标准清晰度电视

SDU (**service data unit**) 业务数据单元

SDV (**switch digital video**) 交换式数字视频广播

SDXC (**SD extended capacity**) 容量扩大化的安全存储卡

SE (**safety engineering**) 安全工程学

sealed building 密闭建筑物

sealed panel 封闭面板

sealing 密封

seamless copper-tube 无缝铜管

search system 系统搜索

seasonal energy efficiency ratio (**SEER**) 季节能源消耗效率

SEC (**security screening**) 安全筛选

SECAM (**sequentiel couleur A memoire**) 塞康制

secluded area 僻静区域

second class power supply 二级市电供电

second degree fault 二级故障

secondary alarm circuit 辅助报警回路

secondary batter 二次蓄电池

secondary circuit 二次电路
secondary dial tone 二次拨号音
secondary distribution 次配线架，二次分配
secondary distribution space 次配线空间
secondary entrance room 次进线间
second-grade low current system worker 二级弱电工
section 截面
section factor 截面系数
sectional ladder 拉梯
sector 扇区
secure digital high capacity (SDHC) SDHC 安全数字大容量卡
secure sockets layer (SSL) 安全套接字层
security 保安
security alarm 安防报警
security alarm system (SAS) 防盗报警系统
security alerting system 安全报警系统
security and protection product 安全防范产品
security and protection system (SPS) 安全防范系统
security assurance 安全保证
security attribute 安全属性
security audit 安全审计
security automation (SA) 安保自动化
security automation system (SAS) 安保自动化系统
security door 安全门
security element 安全要素
security equipment 安全防范设备
security factor 安全系数
security function 安全功能
security function data 安全功能数据
security function policy (SFP) 安全功能策略
security inspection system for anti-explosion 防爆安全检查系统
security level 安全级别
security management 安全管理
security management system (SMS) 安全管理系统
security of information system 信息系统安全
security policy 安全策略
security protection ability 安全保护能力
security screening (SEC) 安全筛选
security subsystem of information system (SSOIS) 信息系统安全子系统
security system 安防系统，安全

系统

security target (ST) 安全目标

security technology prevention system 安全技术防范系统

security-enhanced Linux (SELinux) Linux强制访问控制安全系统

SEER (seasonal energy efficiency ratio) 季节能源消耗效率

segment table (ST) 段表

segmentation 分段

segmentation and reassembly (SAR) 分段与重组

segmentation and reassembly sub-layer (SAR) 分段和重装子层

segregation 隔离

selection information table (SIT) 选择信息表

selective transmit diversity (STD) 选择发射分集

selector valve 选择阀

self phase modulation (SPM) 自相位调制

self test 自测试

self-adapting 自适应

self-adjusting 自校正

self-cleaning 自清洗

self-commutated electronic switch 自换相电子开关

self-coordinating 自协调

self-diagnosing 自诊断

self-extinguishbility 自熄性

self-inferring 自推理

self-learning 自学习

self-organizing 自组织

SELinux (security-enhanced Linux) Linux强制访问控制安全系统

seller 卖方

SEMF (synchronous equipment management function) 同步设备管理功能

semi-finished product (SFP) 半成品

semi-automatic control 半自动控制

semi-flexible coaxial cable 半柔同轴电缆

semi-rigid coaxial cable 半刚同轴电缆

semi-tight buffer 半紧套

sender 发送器

SENECA transducers 电量变送器

sensible cooling capacity 显热制冷量

sensible heat ratio (SHR) 显热比

sensing element 敏感元件

sensing probe 传感探头

sensitive sector 敏感带

sensitivity 灵敏性

sensitivity analysis 灵敏度分析

sensitivity margin 灵敏度冗余

sensitivity to heat 感温灵敏度

sensor 传感器
sensor system 传感器系统
SEP (smart Ethernet protection) 智能以太网保护
separating every pair of wires 线对隔离
separator 分隔板
sequence （视频）序列
sequential contrast 顺序对比度
sequentiel couleur A memoire (SECAM) 塞康制
serial advanced technology attachment (SATA) 串行高级技术附件
serial digital interface (SDI) 串行数字接口
serial line Internet protocol (SLIP) 串行线路因特网协议
serial network 串行网络
serial transmission 串行传输
serial-to-parallel converter/deserializer 串并转换器
serrated roof 锯齿形屋顶
server configuration 服务器配置
service 服务，维护，维修
service access area 维修触及区
service and content protection (SCP) 业务和内容保护
service area 服务区（域）
service area cord 服务区域跳线
service concentration point (SCP) 服务集合[汇集]点
service concentration point cable 服务集合[汇集]点线缆
service data unit (SDU) 业务数据单元
service description table (SDT) 业务描述表
service distribution cable 服务配线线缆
service distribution cabling 分布式服务布缆
service distribution cabling subsystem 分布式服务布缆子系统
service distributor (SD) 服务配线架
service group 服务组
service ID (SID) 业务标识符
service information (SI) 业务信息
service loop 维护盘留
service multiplex and transport 业务复用和传输
service navigation 业务导航
service navigation interface 业务导航接口
service outlet (SO) 服务插座[端口]
service pre-accept 业务预受理
service profile identifier 服务配置文件标识符
service protection 业务保护

service provider (SP) 服务提供商

service system 业务系统

service-level agreement (SLA) 服务等级协议

servo drives 伺服驱动器

SES (smoke extraction system) 排烟系统

session description protocol (SDP) 会话描述协议

session initiation protocol (SIP) 会话启动(初始化)协议,会话发起协议

session manager 会话管理器

set top box of CA security module plug-in CA 安全模块插件机顶盒

set-top box (STB) 机顶盒

seven public nuisances 七种公害

sewage gas 沼气

sewage treatment plant (STP) 污水处理厂

sewerage 下水道系统

SF/FTP 金属编织网、金属箔总屏蔽与金属箔线对屏蔽对绞电缆

SF/UTP 金属编织网与金属箔总屏蔽对绞电缆

SFF (small form factor connector) 小型连接器件

SFN (single frequency network) 单频网

SFN TS distribution network 单频网TS信号分配网络

SFP (security function policy) 安全功能策略

SFP (semi-finished product) 半成品

shading coefficient 遮阳系数

shading coefficient of glass 玻璃遮阳系数

shading coefficient of window (SD) 外窗遮阳系数

shading device 遮阳设备

shadow fading 阴影衰落

shaft horsepower 轴马力[功率]

shaft pump 轴流泵

shaft seal 轴封

shakedown run 试运转

Shannon law 香农定理

shape coefficient 体形系数

shaped steel 型钢

shared IT service 共享 IT 服务

shared office 集中办公

sharp bend 锐弯

sharp freezer 快速[低温]冻结间

sharp freezing 快速[低温]冻结

sharp freezing room 急冻间

sheathed control cable 护套控制电缆

sheet metal 金属片

shelf 机框(无板)

S

shelf-level equipment 货[子]架级设备
shell 外壳
shell and coil condenser 壳管式冷凝器
shell and coil evaporator 壳管式蒸发器
shell and tube condenser 壳管式冷凝器
shell and tube exchanger 壳管式换热器
shell type absorption refrigerating machine 壳式吸收式制冷机
SHF (super high frequency) 超高频
shielded angled patch panel 屏蔽角型配线架
shielded connector 屏蔽连接器
shielded module 屏蔽模块
shielded pair 线对屏蔽
shielded patch panel 屏蔽配线架
shielded RJ45 connector RJ45 屏蔽连接器
shielded twisted pair (STP) 屏蔽双绞线
shielded twisted pair cable (STP cable) 屏蔽双绞线电缆
shielded with dust shutter 防尘盖保护
shielding cabling system 屏蔽布线系统
shock-wave noise 爆音
shop detail drawing 车间加工详图,工厂施工详图
shop drawing 施工图
short circuit 短路
short circuit isolator (SCI) 短路隔离器
short circuit protection (SCP) 短路保护
short message 短消息
short message cell broadcast (SMSCB) 短消息小区广播
short message center (SMC) 短消息中心
short message entity (SME) 短消息实体
short message peer-to-peer (SMPP) 短消息点对点协议
short message service (SMS) 短消息业务
short message service center (SMSC) 短消息服务中心
short message service gateway 短消息服务网关
short trouble 短路故障
short-circuit output current 输出短路电流
show control system 表演[显示]控制系统

SHR (sensible heat ratio) 显热比
shunt valve 旁通阀
shut-off valve 关闭[截止]阀
shutter detector 卷帘门探测器
SI (service information) 业务信息
SI (system integration) 系统集成
SID (service ID) 业务标识符
SID (system identification) 系统识别码
side circuit 单侧电路
side surface 侧面
side view 侧视图
SIF (step index fiber) 突变型光纤,阶跃型折射率光纤
signal 信号
signal alarm 警报器
signal booster 信号增强器
signal controller 信号控制器
signal detection and estimation 信号检测和估计
signal for logical program 逻辑程序信号
signal light 信号灯
signal line 信号线
signal noise ratio (SNR) 信噪比
signal processor 信号处理器
signal relay 信号继电器
signal sustain technology (SST) 信号稳定技术
signal system 7 (SS7) 七号信令系统
signal tracer 信号跟踪器,信号式线路故障寻找器
signaling control channel 信令控制信道
signaling gateway 信令网关
signaling transfer point (STP) 信令转接点
signaling virtual channel (SVC) 信令虚(拟)信道
signal-to-interference ratio (SIR) 信号干扰比
signal-to-noise ratio (SNR) 信号噪声比
signal-to-noise ratio of system 系统信噪比
silent fan 低噪声风机
silica aerogel 二氧化硅气凝胶,带孔硅胶
silica gel 硅胶
silicon 硅脂
silicon control rectifier 可控硅整流器
silicon controlled rectifier (SCR) 可控硅整流器
silicon steel sheet 硅钢片
silicon valley 硅谷(美国地名)
silk screen printing (SSP) 丝网印刷
silvertoun 电缆故障寻迹器

SIM (subscriber identity module card) 用户识别模块卡

simple mail transfer protocol (SMTP) 简单邮件传输协议

simple network management protocol (SNMP) 简单网络管理协议

simple network time protocol (SNTP) 简单网络时间协议

simple object access protocol (SOAP) 简单对象访问协议

simplex 单工

simulation 模拟,仿真

simulation chamber 模拟室

simulation test 模拟试验

simulcrypt 同密

simultaneous contrast 同时对比度

simultaneous factor 同时系数

simultaneous interpretation 同声传译

single acting compressor 单作用压缩机

single cabinet 单机柜

single channel amplifier (for MATV or CATV) (用于共用天线电视或有线电视的)单频道放大器

single channel mode 单声道模式

single column manometer 单管式压力计

single duct air conditioning system 单风道空调系统

single fiber bi-directional 单光纤双向(传输)

single fiber bi-directional transmission 单光纤双向传输

single frequency network (SFN) 单频网络

single leaf damper 单页风口

single mode fiber (SMF) 单模光纤

single phase three wire system 单相三线制

single port faceplate 单口面板

single time 单次

single UPS 单台 UPS

single vane rotary compressor 单叶回转式压缩机

single-band infrared flame detector 单波段红外火焰探测器

single-input multiband amplifier 单输入多频段放大器

single-line digital subscriber line (SDSL) 单线(路)数字用户线路

single-mode 单模(光纤)

single-mode fiber (SMF) 单模光纤

single-mode fibre connector 单模光纤连接器

single-mode optical fiber (SMF) 单模光纤

single-mode optical fiber jumper wire 单模光纤跳线

single-pole double-throw (SPDT) 单刀双掷开关
single-program transport stream (SPTS) 单节目传输流
single-shaping 一次成型
SIP (session initiation protocol) 会话启动(初始化)协议[会话发起协议]
SIP URI 会话初始协议 URI
siphon action 虹吸作用
siphon barometer 虹吸式气压计
siphonage 虹吸
SIR (signal-to-interference ratio) 信号干扰比
SIT (selection information table) 选择信息表
site commissioning 现场调试
site construction application report 进场施工申请报告
site contact 现场联系人
site inspection 工地勘察
site regulation 现场监管
site training 现场培训
skin effect 趋肤效应
SLA (service-level agreement) 服务等级协议
slab insulant 板状绝缘材料
slag cotton 矿渣棉
slag pool 渣池
slag tapping boiler 液态排渣锅炉
slag wool 矿渣棉
slave clock 子钟,从时钟
SLD (straight line distance) 直线距离
sleeve expansion joint 套筒伸缩器
sleeve for duct passing through wall 风管穿墙用套管
slice 条带
sliding dust shutter 滑盖式防尘盖
sliding rail 滑动轨
slight salt fog 轻度盐雾
SLIP (serial line Internet protocol) 串行线路因特网协议
slot position 槽位
slow associated control channel (SACCH) 慢速随路控制信道
slow flashing 慢闪
slow frequency hopped multiple access 慢跳频多址接入
SLS (safety limit switch) 保险总开关
SM (synchronous multiplexer) 同步复用器
small exchange configuration 小交换机配置,小型交换局配置
small form factor connector (SFF) 小型连接器件
small impairment 小损伤
small office/home office (SOHO) 小型公寓式办公室或家庭办公室

S

smart card 智能卡

smart Ethernet protection (SEP) 智能以太网保护

smart grid 智能电网

smart home 智能家居

smart home application 智能家居应用

smart home platform 智能家居业务平台

smart home system 智能家居系统

smart home terminal 智能家居终端

smart house 智慧屋,智能家居

smart house technology 智能家居技术

smart hybrid terminal 智能融合终端

smart meter 智能电表[仪表]

smart sensor 智能传感器

smart transducer 智能变送器

smart wiring box 智能布线箱

SMB (surface mount box) 表面安装盒

SMC (short message center) 短消息中心

SME (short message entity) 短消息实体

smearing 拖尾效应

smearing time 拖尾时间

SMF (single-mode fiber) 单模光纤

SMF (single-mode optical fiber) 单模光纤

SMI (structure of management information) 管理信息结构

smoke alarm 烟雾报警器

smoke blower 排烟机

smoke compartment 防烟分区

smoke curtain 挡烟垂壁

smoke damper 排烟阀

smoke density 烟密度

smoke extraction system (SES) 排烟系统

smoke fire damper 排烟防火阀

smoke fire detector 感烟火灾探测器

smoke proof staircase 防烟楼梯

smoke protection system 防烟系统

smoke/fire detection 烟雾火灾检测

smoke-free fire 无烟火灾

smoke-proof staircase 防烟楼梯间

smoking room 吸烟室

smoldering 阴燃

smoothness 平整度

smothering 窒息,断氧灭火

smouldering 阴燃

SMP (symmetry multiprocessors) 对称多处理机

SMPP (short message peer-to-peer) 短消息点对点协议

SMPTE (the Society of Motion Picture and Television Engineers) 电影与电视工程师协会
SMR (specialized mobile radio) 专用移动无线通信
SMS (security management system) 安全管理系统
SMS (short message service) 短消息业务
SMSC (short message service center) 短消息服务中心
SMSCB (short message cell broadcast) 短消息小区广播
SMTP (simple mail transfer protocol) 简单邮件传输协议
SNA (system network architecture) 系统网络结构
snap-in module 插入式模块
SNG (satellite news gathering) 卫星新闻采集
SNMP (simple network management protocol) 简单网络管理协议
snow and ice 冰雪
SNR (signal noise ratio) 信噪比
SNR (signal-to-noise ratio) 信号噪声比
SNTP (simple network time protocol) 简单网络时间协议
SO (service outlet) 服务插座[端口]
SOAP (simple object access protocol) 简单对象访问协议
SoC (system-on-a-chip) 片上系统
socket 插座
socket unshielded 非屏蔽插座
soft ground 软接地
soft handoff 软切换
soft starter 软启动器
software architecture 软件体系结构
software development kit (SDK) 软件开发工具包
software license 软件许可证
software product specification (SPS) 软件产品规格说明
SOHO (small office/home office) 小型公寓楼办公室或家庭办公室
solarization 曝光过度
solder 焊料
solder wire 焊锡丝
soldering flake 焊片
solenoid valve 电磁阀
solid conductor 实心导体
solid copper wire 实芯铜线
SONET (synchronous optical network) 同步光网络
SONY/PHILIPS digital interface format (S/PDIF) SONY、PHILIPS 数字接口格式
SOS (safety observation station) 安

S

全观察站

sound and image synchronization 声像同步

sound and light alarm 声光警报器

sound carrier 伴音载波

sound carrier frequency 伴音载频

sound channel 声道

sound delay of video conference 视频会议的声音延时

sound field 声场

sound field irregularity 声场不均匀度

sound focusing 声聚焦

sound on sync 声同步

sound pressure level (SPL) 声压级

sound reduction index 降噪指数，声衰减指数

sound reinforcement system 扩声系统

sound signal 声音信号

source 信号源

Southern Standard Building Code (SSBC) 南方标准建筑规范

SP (service provider) 服务提供商

space diversity 空间分集

space division multiple access (SDMA) 空分多址

space efficient 空间效率

space time block coding (STBC) 空时分组编码

spanning tree protocol (STP) 生成树协议

spare part 备件

SPC exchange 程控交换机

SPD (surge protective device) 浪涌保护器

SPDT (single-pole double-throw) 单刀双掷开关

speaker 扬声器

speakers distribution 广播扬声器布点

speakers, A/V & home theater 家庭影院系统

special test software 专用测试软件

special test system 专用测试系统

special tool 专用工具

specialized mobile radio (SMR) 专用移动无线通信

specification 规范[规格]书

specification description language (SDL) 规范描述语言

specifier 规划方

specimen 样品

specimen page 样张

spectral shape 光谱形状

spectrum spreading 扩频，频谱扩展

speculum 反射镜

speech coding 语音编码

speech communication 话音通信

speech interface 话音接口

speech transmission index of public address (STIPA) 扩声系统语言传输指数

SPL (sound pressure level) 声压级

SPL (split charging) 分摊计费

splice 接合

splice closure for optical cable 光纤连接器

splice closure for outdoor optical cable 室外光缆连接器

splice connection 熔纤连接

splice for optical fiber 光纤接头

splice holder 熔接固定

splice protection 熔接保护

splice tray 熔纤盘

splice tray with cover 带盖熔纤盘

split charging (SPL) 分摊计费

splitter 分支[配]器

SPM 自相位调制

SPM (self phase modulation) 自相位调制

spontaneous ignition 自燃

spotlight 聚光灯

spot-type detector 点型探测器

spot-type fire detector 点型火灾探测器

spray 喷雾

spray fire pump 喷淋消防泵

spread rate 传播速率

spread spectrum (SS) 扩展频谱

spring-return actuator 弹簧复位启动装置

sprinkler alarm valve water motor alarm 报警阀水警铃

sprinkler system 自动喷水灭火系统

SPS (secondary power supply) 二次电源

SPS (security and protection system) 安全防范系统

SPS (software product specification) 软件产品规格说明

SPTS (single-program transport stream) 单节目传输流

SQL (structured query language) 结构化查询语言

squelch circuit 静噪电路

squelch opening and closing level 静噪开启电平和闭锁电平

SRAM (static random access memory) 静态随机存取存储器

SRC (sample rate conversion) 采样频率转换

SRL (structural return loss) 结构回波损耗

SRS (stimulated Raman scattering) 受激拉曼散射

SRV (safety relief valve) 安全减压阀

SS (spread spectrum) 扩展频谱

SS (supplementary service) 补充业务

SS7 (signal system 7) 七号信令系统

SSBC (Southern Standard Building Code) 南方标准建筑规范

SSC (SSF scope of control) SSF 控制范围

SSF (SSOIS security function) SSOIS 安全功能

SSF scope of control (SSC) SSF 控制范围

SSL (secure sockets layer) 安全套接字层

SSOIS (security subsystem of information system) 信息系统安全子系统

SSOIS security function (SSF) SSOIS 安全功能

SSOIS security management SSOIS 安全管理

SSOIS security policy (SSP) SSOIS 安全策略

SSP (silk screen printing) 丝网印刷

SSP (SSOIS security policy) SSOIS 安全策略

SST (signal sustain technology) 信号稳定技术

ST (security target) 安全目标

ST (segment cable) 段表

ST connector ST(光纤)连接器

stabilizing element from pairs 十字隔离

stack HUB 堆叠型集线器

stacking 堆垛

stage 舞台

stainless steel 不锈钢

stair 楼梯

stair landing 楼梯平台

stairway 楼梯

standalone dedicated control channel (SDCCH) 独立(专用)控制信道

stand-alone electrical fire monitoring detector 独立式电气火灾监控探测器

standalone server 独立服务器

standard (DIN) rail 标准(DIN)导轨

standard 19 inches cabinet 标准 19 英寸机柜

standard atmospheric condition 标准大气条件

standard building code (SBC) 标准建筑规范

standard definition barcode image 标清码图

standard definition television

(SDTV) 标准清晰度电视
standard socket 标准插座
standard white 标准白
standardization 标准化
standardized management 规范管理
standby power 备用电源
standby redundant UPS 备用冗余 UPS
standing display screen 落地显示屏
standing wave 驻波
star network 星型网络
start code 起始码
static bypass (electronic bypass) 静态旁路(电子旁路)
static data 静态数据
static load-bearing capacity 静态承重
static pressure 静压
static random access memory (SRAM) 静态随机存取存储器
static shielding 静电屏蔽
static state condition 静态条件
stationary equipment 静置[固定]设备
stationary use 固定使用
statistical analysis system (SAS) 统计分析系统
statistical multiplexing 统计复用
statistical time division duplex (STDD) 统计时分复用
STB (set-top box) 机顶盒
STBC (space time block coding) 空时分组编码
STD (selective transmit diversity) 选择发射分集
STDD (statistical time division duplex) 统计时分复用
STDM (synchronous time division multiplexer) 同步时分复用器
STDM (synchronous time division multiplexing) 同步时分复用
steel plastic pipe 钢性塑料管
steel structure of display screen 显示屏钢构架
steel-made cable support system 钢质电缆桥架
stent 支架
step index fiber (SIF) 突变型光纤,阶跃型折射率光纤
stepping 步进
stereo mixing 立体声混合
stereo simulation 立体声模拟
stereophony 立体声
still picture 静止图像
still store 静止存储器
stimulated Brillouin scattering (SBS) 受激布里渊散射
stimulated Raman scattering (SRS)

受激拉曼散射

STIPA (speech transmission index of public address) 扩声系统语言传输指数

STM (synchronous transport module) 同步传送模块

STN (super twisted nematic) 超扭曲向列

STN (switched telephone network) 电话交换网络

stop condition 停机条件

storage area network (SAN) 存储区域网络

storage cell 蓄电池,计算机存储单元

storage facility 储存设施

storage multimedia 存储多媒体

storage of material 材料存储

stored energy time 储能供电时间

STP (sewage treatment plant) 污水处理厂

STP (shielded twisted pair) 屏蔽双绞线

STP (signaling transfer point) 信令转接点

STP (spanning tree protocol) 生成树协议

STP cable (shielded twisted pair cable) 屏蔽双绞线电缆

straight joint 直通接头

straight line distance (SLD) 直线距离

strain relief 应力消除

stranded conductor 多芯绞合导体,绞股导体,绞合的铜丝

stranded copper wire 绞铜线

stranded loose tube optical fibre cable 层绞式光缆

streaking 拖尾

streaming media 流媒体

stripped back 剥离

structural element 结构化部件

structural fire protection 建筑消防

structural return loss (SRL) 结构回波损耗

structural steel 结构钢架

structural stress 结构应力

structure 结构

structure chart 结构图

structure of management information (SMI) 管理信息结构

structured cabling system (SCS) 结构化布线系统

structured premises cabling 结构化布线

structured query language (SQL) 结构化查询语言

stuffing bits 填充位,位填充

SU (subscriber unit) 用户单元

sub-station 副分机
sub-contractor 分包商
subcontractor for equipment installation 设备安装分包商
subnetting 子网划分
subscriber access point (SAP) 用户接入点
subscriber authorization system (SAS) 订户[用户]授权系统
subscriber connector (SC) 用户连接器
subscriber identity module card (SIM) 用户识别模块卡
subscriber optical cable 用户光缆
subscriber unit (SU) 用户单元
subscribers feeder 用户线
subsystem 子系统
subtitle 字幕
suite of installation parts 安装成套件
super high frequency (SHF) 超高频
super twisted nematic (STN) 超扭曲向列
supervise 监理
supervisor call interrupt 监控程序请求中断,访管中断
supervisory channel 监控信道
supervisory computer 上位机
supervisory computer control (SCC) 监督计算机控制
supervisory computer control system (SCCS) 计算机监控系统
supervisory control 监督控制
supervisory network engine 网络监控引擎
supervisory system 监控系统
supplementary channel 增补频道
supplementary service (SS) 补充业务
supplier 供应商
supply air (SA) 送风
supply impedance 电源阻抗
support area 支持区
supporting basic installation 配套基础设施
surface 表面
surface guarding 面警戒
surface mount box (SMB) 表面安装盒
surface-mounted 表层安装,明装
surface-mounted installation 表面安装
surface plastic spray 表面喷塑
surface resistance 表面电阻
surface resistivity 表面电阻率
surface treatment of material 材料表面处理
surge protective device (SPD) 浪涌保护器

surround sound 环绕声
surrounding 环境
surveillance 监视
surveillance & control center 监控管理中心
surveillance and control center 监控管理中心
surveillance area 监视区
surveillance camera 监控摄像机
surveillant 监视者
sustainability 可持续性
SVC (scalable video coding) 可伸缩视频编码
SVC (signaling virtual channel) 信令虚(拟)信道
SVP (saturated vapor pressure) 饱和蒸气压
SW (switch) 交换机,开关,选择
switch (SW) 交换机,开关,选择
switch digital video (SDV) 交换式数字视频广播
switch type 开关型号
switched telephone network (STN) 电话交换网络
switching matrix 交换矩阵
symmetrical cable 对称电缆
symmetrical connection 对称连接
symmetrical pair/quad cable 对绞或星绞对称电缆
symmetry multiprocessors (SMP) 对称多处理机
sync restoration 同步恢复
sync stripping 同步剥离
synchronization 同步
synchronization channel 同步信道
synchronization signal unit (SYU) 同步信号单元
synchronous communication 同步通信
synchronous digital hierarchy (SDH) 同步数字分层结构
synchronous equipment management function (SEMF) 同步设备管理功能
synchronous multiplexer (SM) 同步复用器
synchronous optical network (SONET) 同步光网络
synchronous time division multiplexer (STDM) 同步时分复用器
synchronous time division multiplexing (STDM) 同步时分复用
synchronous transfer 同步传输
synchronous transport module (STM) 同步传送模块
synoptic chart 天气图
syntax element 语法元素
syslog 系统日志
system 系统
system acceptance 系统验收

system acceptance test (SAT) 系统验收测试

system accessories 设备安装辅料

system chart 系统框图

system checking and measuring 系统检查和测量

system circuit integrity 系统线路完整性

system component 系统组件

system configuration 系统配置

system deficiency liability period 系统缺陷责任期

system drawing 系统图

system identification (SID) 系统识别码

system integration (SI) 系统集成

system maintenance 系统维护

system maintenance valve 系统检修阀

system management 系统管理

system network architecture (SNA) 系统网络结构

system service 系统服务

system set-up diagram 系统配置图

system-on-a-chip (SoC) 片上系统

SYSTIMAX SYSTIMAX 综合布线系统(美国康普公司的综合布线系统品牌名之一)

SYU (synchronization signal unit) 同步信号单元

SYV 实心聚乙烯绝缘聚氯乙烯护套同轴射频电缆

T

T568 T 568 打线规则

T568A T568A 类连接方式

T568B T568B 类连接方式

TAC Vista system 施耐德 TAC Vista 楼控系统

tachometer 流速计

TACS (total access communication system) 全接入通信系统

tag 标签

tail fiber 尾纤

tailoring 裁剪，拆条

tailor-made 定制

tall building 高层建筑

tamper detector 防瞎摆弄探测器(消防)

tamper device 防拆功能

tank sensor 水箱传感器

tape cable 带状电缆

target of classified security 等级保护对象

target surface 靶面

task illumination 工作照明

TBIC (total building integration cabling) 整体楼宇集成布线(美国西蒙公司的综合布线品牌解决方案名称)

TC (temperature coefficient) 温度系数

TCH (traffic channel) 业务流量信道

TCH/H (half rate traffic channel) 半速率业务信道

TCL (transverse conversion loss) 横向转换损耗

TCM (trellis coded modulation) 格栅编码调制

TCO (total cost of ownership) 所有权费用总额

TCP (transmission control protocol) 传输控制协议

TCP offload engine (TOE) TCP 卸载引擎

TCP/IP (transmission control protocol/Internet protocol) 传输控制协议/因特网互联协议

TCS (transmission convergence

sublayer) 传输会聚子层

TCTL (transverse conversion transfer loss) 横向转换传送损耗

TCW (tinned copper wire) 镀锡铜丝

TDM (time division multiplex) 时分复用

TDR (time domain reflectometer) 时域反射

TDT (time and date table) 时间和日期表

TE (terminal equipment) 终端设备

teaching building 教学楼

tearing 图像撕裂

technical coordination 技术协调

technical description 技术性描述

technical disclosure 技术交底

technical liaison 技术联络

technical office protocol (TOP) 技术办公系统协议

technical protection 技术防范

technical report (TR) 技术报告

technical service 技术服务

technical specification 技术规范

technical specification group 技术规范组

technical worker 技术工人

technological management 工艺管理

TEE (trusted execution environment) 可信执行环境

telautogram 传真电报

telechirics 遥控系统

telecine 电视电影机

telecommunications equipment room (TER) 电信设备室,通信机房

telecommunications junction box for home 住宅信息配线箱

telecommunications network voltage circuit 电信网络电压线路

telecommunications 电信

telecommunications cable 电信线缆

telecommunications conduit 电信管道

telecommunications equipment 电信设备

Telecommunications Industry Association (TIA) 电信工业协会(美国)

telecommunications infrastructure 电信基础设施

telecommunications non-central room 电信非中心机房

telecommunications outlet (TO) 电信插座,信息点,工作区

telecommunications room 电信间

Telecommunications Technology Committee (TTC) 电信技术委

员会

telegraph circuit 电报电路

telegraph fire alarm system 电报消防报警系统，编码火灾报警系统

telephone alarm device 电话报警设备

telephone call 电话报警

telephone cord 电话软线

telephone exchange 电话交换台

telephone frequency 话频

telephone jack 电话插孔

telephone line 电话线

telephone network 电话网

telephone set 电话机

telephone system (second generation) 第二代电话系统

telephone trunk 电话中继线

teleprinter 电传打字机

tele-prompter 字母提示机

telescopic slide 伸缩滑轨

teletext 图文电视

teletype circuit 电传打字机电路

television camera 电视摄像机

television monitoring equipment 电视监控设备

television operating system (TVOS) 智能电视操作系统

television studio 电视演播室

television transmitting tower 电视发射塔

temperature cable 感温光缆

temperature coefficient (TC) 温度系数

temperature controller 温度控制器

temperature fire detector 感温火灾探测器

temperature gauge 温度测试仪

temperature indicating equipment 温度指示设备

temperature inversion 逆温

temperature measuring probe 测温器

temperature rise 温升

temperature scale 温标

temperature transducer 温度传感器

temporary enclosure of building 建筑物的临时围护

temporary high service system 临时高压系统

temporary mobile station identity (TMSI) 临时移动台标识

temporary safe refuge (TSR) 临时安全避难所

temporary structure 临时性建筑

tendering document 招标文件

tendering product 招标产品

tendering specification 招标规

格书

tenement 公寓

tensile performance 拉伸［抗张］性能

tensile strength 拉伸强度

tensioned cable 张拉索

TER（telecommunications equipment room） 电信设备室

TERA connector TERA 连接器

terminal block 接线盒

terminal board 接线端子板

terminal box 终端盒

terminal box for optical cable 光缆终端分线盒

terminal building 航站楼

terminal control unit 终端控制单元

terminal equipment（TE） 终端设备

terminal lug 接线端子

terminal of IoT 物联网终端

terminal unit 终端装置

terminating quality 端接质量

terminating resistor 终端［端接］电阻器

termination 端接

termination multiplexer 终端复用器

termination point 端接点

termination switch 端接开关

terminology 术语

test expense 测试费

test form 测试表格

test interface（TI） 测试接口

test load 测试负载

test model 测试模型

test of no load condition 空载工况试验

test picture 测试图像

test record 测试记录

test report 测试报告

test signal generator（TSG） 测试信号发生器

testing 测试

TF card（trans-flash card） TF 卡

TFT（thin film transistor） 薄膜晶体管

TFTP（trivial file-transfer protocol） 简单文件传送协议

thatch roof 茅草屋顶

the Society of Motion Picture and Television Engineers（SMPTE） 电影与电视工程师协会

theatrical stage 剧场舞台

theodolite 经纬仪

thermal baffle 隔热板

thermal barrier 隔热层

thermal cut-off 隔热挡板

thermal detector 感温探测器

thermal device 过热保护装置

thermal graphic 热图像

thermal imaging camera (TIC) 热成像摄像机

thermal protection shield 隔热罩

thermal radiation 热辐射

thermal radiation shield 热辐射防护屏

thermal sensor 热传感器

thermal shield 隔热罩

thermal shield material 隔热材料

thermal-protection material (TPM) 隔热材料

thermal-protective coating (TPC) 隔热层

thermistor 热敏电阻

thermo element 热电元件

thermocouple 热电偶

thermodynamic diagram 热力图

thermoelectric effect 温差电效应

thermoelectric effect detector 热电效应探测器

thermoelectric sensor 热电传感器

thermograph 温度记录器

thermometer 温度计

thermoplastic duct 热塑性塑料管

thermosensitive element 热敏元件

thermostat 恒温(调节)器

thermostats & HVAC controls 恒温器与采暖通风空调控制

thermoweld technique 熔接工艺

thick coaxial cable 粗同轴电缆

thickness 厚度

thin film transistor (TFT) 薄膜晶体管

third degree fault 三级故障

third-party test institution 第三方测试机构

thread Neill-Concelman (TNC) 螺纹连接器

three connections 三连接

three level low current system worker 三级弱电工

three way 三通

three-phase electronic power analyzing electric energy meter 三相电力分析能量表[计]

threshold value 阈值

through-penetrant 墙身贯穿物

through-penetration 墙身贯穿洞孔

throw distance 投影距离

thunderstorm day 雷暴日

TI (test interface) 测试接口

TIA (Telecommunications Industry Association) 电信工业协会(美国)

TIC (thermal imaging camera) 热成像摄像机

tie cable 连接线缆

tie line 专线

tie switch 互连开关

tie trunk 连接中继线
tilt angle 倾斜角
tilt sensor 倾斜探测器
timber construction 木结构
timbre 音质
time and date table (TDT) 时间和日期表
time division multiplex (TDM) 时分复用
time domain reflectometer (TDR) 时域反射
time line 时间线
time of fire alarming 火灾报警时间
time offset table (TOT) 时间偏移表
time-saving installation 省时安装，快速安装
tinned braided copper 丝网编织镀锡铜丝
tinned copper wire (TCW) 镀锡铜丝
tint 色调
TLS (transport layer security) 传输层安全(协议)
TMDS (transition minimized differential signaling) 跃变最小化差分信号
TMSI (temporary mobile station identity) 临时移动台标识
TMY (typical meteorological year) 典型气象年
TNC (thread Neill-Concelman) 螺纹连接器
TO (telecommunications outlet) 电信插座,信息点,工作区
TOE (TCP offload engine) TCP卸载引擎
toeboard 趾板
toggle 套环
token passing (TP) 令牌传递
tolerance 容错
tolerance band 公差[允差]带
tone 音色
tone control 音调控制
tone generator 音频发生器
tool-less 免工具
TOP (technical office protocol) 技术办公系统协议
top exit 顶部出口
top floor 顶层
top level ventilation 顶层通风
topology 拓扑结构
topology of network 网络拓扑
torsion 扭转
TOT (time offset table) 时间偏移表
total access communication system (TACS) 全接入通信系统
total building integration cabling

(TBIC) 整体楼宇集成布线(美国西蒙公司的综合布线品牌解决方案名称)

total cost of ownership (TCO) 所有权费用总额

total dispatch room 总调度室

total distortion factor 总失真系数

total harmonic distortion 总谐波失真

total quality control (TQC) 全面质量管理

total shading coefficient of window 窗的总遮阳系数,外窗综合遮阳系数

total UPS transfer time UPS 总切换时间

TO-type interface TO 类型接口

tower cupola 瞭望塔

tower structure 塔式结构

towerman 瞭望员

toxicity 毒性

TP (token passing) 令牌传递

TP (transition point) 转接点[处]

TPC (thermal-protective coating) 隔热层

TPM (thermal-protection material) 隔热材料

TP-PMD (twisted pair physical medium dependent) 双绞线物理层相关媒体

TPS (transmission parameter signaling) 传输参数信令

TQC (total quality control) 全面质量管理

TR (technical report) 技术报告

trade off 权衡判断

trademark right 商标权

traffic channel (TCH) 业务流量信道

traffic charge equipment 交通收费设备

traffic detector 交通检测设备

traffic management equipment 交通管理设备

traffic system software 交通系统软件

trailer coach 活动房

trailer court 活动房屋停放场

trailer park 活动房屋停放场

training material 培训资料

training room 培训教室

trans-coding 转码

transfer impedance 转移阻抗

transfer switch 转换开关

transfer time 切换时间

trans-flash card (TF card) TF 卡

transform coefficient 变换系数

transient 瞬态

transient current 瞬时电流

transient over pressure 瞬时过

电压

transistor 晶体管

transition minimized differential signaling (TMDS) 跃变最小化差分信号

transition point (TP) 转接点[处]

transmission circuit 传输电路

transmission control protocol (TCP) 传输控制协议

transmission control protocol/Internet protocol (TCP/IP) 传输控制协议/因特网互联协议

transmission convergence sublayer (TCS) 传输会聚子层

transmission distance 传输距离

transmission equipment 传输设备

transmission frequency respond 传输频率响应

transmission line 传输线路

transmission link for live broadcasting 直播传输通路

transmission matrix 传输矩阵

transmission media 传输介质

transmission mode 传输模式

transmission multimedia 传输多媒体

transmission of consumption 传输消耗

transmission parameter signaling (TPS) 传输参数信令

transmission performance 传输性能

transmission window 传输窗口

transmit (XMT) 发送，发射，传输，传播

transmit power accuracy 发送电平精度

transmit power step size 发送电平步长

transmit-receive unit (TRU) 发射接收装置

transmitter (TX) 发(送)机，传送[传讯]器，发射机

transmitter automation unit 发射机自动化单元

transmitter conducted spurious 发送机寄生产物

transmitter frequencies 发送频率

transmitter frequency accuracy 发送机频率精度

transmitter frequency step size 发送机输出频率步长

transmitter front porch time 发送机前沿时间

transmitter out-of-band noise suppression 发送带外噪声抑制

transmitter power delta between "mark" and "space" 传号和空号的发送机功率差

transmitter slew rate 发送机数位转换时间

T

transmitter supervision system 发射机监控系统,发射机运行管理系统
transmitting antenna 发射天线
transmitting equipment 发射设备
transmitting station operation management system 传送站运行管理系统
transparent clock 透明时钟
transparent interconnection of lots of links (TRILL) 多链路透明互联
transponder 异频雷达收发机,应答器
transport layer 传输层
transport layer security (TLS) 传输层安全(协议)
transport stream (TS) 传送流
transport stream ID (TSID) 传送流标识符
transport stream packet header 传送流包头
transportation house 运输用建筑物
transportation vehicle 交通工具
transverse conversion loss (TCL) 横向转换损耗
transverse conversion transfer loss (TCTL) 横向转换传送损耗
trapezoid correction 梯形校正
tray 托盘
treatment location 诊疗区域
trellis coded modulation (TCM) 格栅编码调制
triac 三端双向可控硅开关元件
triac output 双向可控硅输出
trial operation 操作检测
trial running 试运行
trigger circuit 触发电路
trigger device 触发装置
trigger output 触发输出
trigger signal 触发信号
tri-level sync 三级同步
TRILL (transparent interconnection of lots of links) 多链路透明互联
triple play 三网融合
trivial file transfer protocol (TFTP) 简单文件传送协议
troposphere 对流层
troubleshooting 故障处理
TRS (trunked radio system) 集群无线[移动]通信系统
TRU (transmit-receive unit) 发射接收装置
true color 真彩色
true video-on-demand (TVOD) 视频实时点播
trunk 主干[中继]线,长途线
trunk amplifier 干线放大器
trunk transmission 干线传输

trunked radio system (TRS) 集群无线[移动]通信系统

trunking communication 集群通信

trunking system for mobile radio 集群移动无线电通信

trusted channel 可信信道

trusted execution environment (TEE) 可信执行环境

trusted path 可信路径

TS (transport stream) 传送流

TS transport jitter TS流传输抖动

TSG (test signal generator) 测试信号发生器

TSID (transport stream ID) 传送流标识符

TSR (temporary safe refuge) 临时安全避难所

TTC (Telecommunications Technology Committee) 电信技术委员会

tunnel 隧道

turbine flowmeter 涡轮式流量计

TV conference 电视会议

TV line (TVL) 电视线

TV modulator 电视调制器

TV screen 电视墙

TVL (TV line) 电视线

TVOD (true video-on-demand) 视频实时点播

TVOS (television operating system) 智能电视操作系统

twin lead 平行馈线

twin-axial cable 双芯同轴电缆

twining cable 双绞电缆

twist 扭绞

twisted pair 对[双]绞线

twisted pair cable 双绞线电缆

twisted pair physical medium dependent (TP-PMD) 双绞线物理层相关媒体

twisted wire 双扭线

two speed motor 双速电机

two-way telephone 对讲电话

TX (transmitter) 光发(送)机，信号传送器，发射机

type 类型

type A generic cable A类布缆

type B generic cable B类布缆

type test 型式检验

typical measured value 典型测量值

typical meteorological year (TMY) 典型气象年

U

U (unit) （标准机柜或机架）设备安装尺寸单位

U/FTP 金属箔线对屏蔽对绞电缆

U/FTQ 金属箔线对屏蔽四芯星绞电缆

U/UTP 非屏蔽对绞电缆

U/UTQ 非屏蔽八芯星绞电缆

U2U signaling (user-to-user signaling) 用户到用户信令

UA (user agent) 用户代理

UART (universal asynchronous receiver/transmitter) 通用异步收发器

UB (unique browser) 独立访问者

UCC (Uniform Code Council) 统一代码委员会（美国）

UCS (universal character set) 通用字符集，字符系统

UDP (user datagram protocol) 用户数据报协议

UHDTV (ultra-high-definition television) 超高清电视

UHF (ultra-high frequency) 特高频

UI (unit interval) 单位时间间隔

UI (user interface) 用户交互界面，用户接口

UID (user identification) 用户识别码，用户标识

UL (Underwriter Laboratories Inc.) 保险商试验所（美国）

UL (uplink) 上行链路

ultra-extended graphics array (UXGA) 超扩展图像格式

ultra-high frequency (UHF) 特高频

ultra-high-definition television (UHDTV) 超高清电视

ultra-physical contact (UPC) 弧形接触面

ultrafine coaxial cable 极细同轴电缆

ultrasonic 超声波

ultrasonic alarm 超声波报警器

ultrasonic cleaning machine 超声波清洗机

ultrasonic sensor 超声波传感器
unattended room 无人值守机房
unauthorized access 非法入侵
unbalance attenuation near end 近端不平衡衰减
unbalanced attenuation 不平衡衰减
unbalanced line 不平衡线路
unbalanced load 不平衡负载
unbalanced signal 不平衡信号
unbalanced transmission 不平衡传输
under water cable (UWC) 水下线缆
underground pipe 地下管道
underground railroad 地铁
underground space 地下空间
undervoltage 欠压
underwater tunnel 水底隧道
Underwriter Laboratories Inc. (UL) 保险商试验所(美国)
undetachable 不可拆卸
unicast 单播
unicast reverse path forwarding (URPF) 单播逆向路径转发
Unicode 统一码,单一码(通用字符编码标准)
unidirectional air flow clean room 单向流洁净室
unidirectional broadcast 单向广播
Uniform Code Council (UCC) 统一代码委员会(美国)
uninterruptible power supply (UPS) 不间断电源
unique browser (UB) 独立访问者
unit (U) (标准机柜或机架)设备安装尺寸单位
unit interval (UI) 单位时间间隔
unit under test 被测单元
unitary air-conditioners for computer and data processing room 计算机与数据处理室一体式空调器,机房用单元式空气调节机
universal asynchronous receiver/transmitter (UART) 通用异步收发器
universal character set (UCS) 字符系统,通用字符集
universal faceplate 通用型面板
universal personal telecommunication (UPT) 通用个人电信
universal plug and play (UPnP) 通用即插即用
universal product code (UPC) 通用产品码
universal PTZ camera 万向云台摄像机
universal serial bus (USB) 通用串行总线
UNIX UNIX 操作系统

U

unkeyed socket 非连锁插座
unscreened balanced cable 非屏蔽对绞电缆
unscreened cable 非屏蔽电缆
unscreened connector 非屏蔽连接器
unshielded cable 非屏蔽电缆
unshielded distribution frame 非屏蔽配线架
unshielded RJ45 connector 非屏蔽RJ45连接器
unshielded RJ45 jumper wire 非屏蔽RJ45跳线
unshielded RJ45 module 非屏蔽RJ45模块
unshilded twisted pair (UTP) 非屏蔽双绞线
untwisted length 解开扭绞长度
unvisualable intercom 非可视对讲机
UPC (ultra-physical contact) 弧形接触面
UPC (universal product code) 通用产品码
upgrade of the SIM card SIM卡升级
uplink (UL) 上行链路
UPnP (universal plug and play) 通用即插即用
upper/lower alarm limit 告警上下限
UPS (uninterruptible power supply) 不间断电源
UPS double conversion UPS双变换
UPS double conversion with bypass 带旁路UPS双变换
UPS efficiency UPS效率
UPS functional unit UPS功能单元
UPS inrush current UPS冲击电流
UPS interrupter UPS断路器
UPS isolation switch UPS隔离开关
UPS line interactive operation UPS互动运行
UPS line interactive operation with bypass 带旁路的UPS互动运行
UPS maintenance bypass switch UPS维修旁路开关
UPS maximum input current UPS最大输入电流
UPS passive standby operation UPS后备运行
UPS rated input current UPS额定输入电流
UPS rectifier UPS整流器
UPS switch UPS开关
UPS switch operation UPS开关操作

UPS unit UPS单元

upscaling 倍线

UPT (universal personal telecommunication) 通用个人电信

UPVC pipe UPVC管

urban cells 城市单元,城市蜂窝

urban road tunnel 城市道路隧道

urgency signal 紧急信号

URPF (unicast reverse path forwarding) 单播逆向路径转发

USB (universal serial bus) 通用串行总线

USB communication line USB通信线

user agent (UA) 用户代理

user authentication 用户鉴别

user datagram protocol (UDP) 用户数据报协议

user distribution network 用户分配网

user identification (UID) 用户识别码,用户标识

user information transmission device 用户信息传输装置

user interface (UI) 用户界面

user interface profile 用户接口档案

user junction box for home 家用配线箱

user manual 使用说明书,使用手册

user profile 用户配置,用户轮廓文件

user service identity module (USIM) 用户业务识别模块

user terminal 用户终端

user view 用户视图

user's guide 使用指南

user-subject binding 用户-主体绑定

user-to-user signaling (U2U signaling) 用户到用户信令

USIM (user service identity module) 用户业务识别模块

UTC (coordinated universal time) 协调通用(世界)标准时间

UTP (unshilded twisted pair) 非屏蔽双绞线

UTP network port UTP网络接口

UV flame detector 紫外火焰探测器

UV resistance 抗紫外线辐射

UV-proof 防紫外线

UWC (under water cable) 水下线缆

UXGA (ultra-extended graphics array) UXGA图像格式

V

V.V.V. (valid voltage value) 电压有效值

v-coeff v-coeff 符号(代表线缆衰减的温度系数)

valid user 有效用户

valid voltage value (V. V. V.) 电压有效值

value-added service (VAS) 增值服务[业务]

valve 阀

valve regulated sealed (secondary) cell 阀控密封(二次)蓄电池

VANC (vertical ancillary data) 垂直辅助数据

vandal resistant 防破坏(护罩)

variable bit rate (VBR) 变码率,可变比特率

variable distribution system 变配电系统

variable length code (VLC) 可变长度码

variable resistance (VR) 可变电阻

variable-frequency drive (VFD) 变频器

VAS (value-added service) 增值服务[业务]

VAT (video analysis technology) 视频分析技术

VAV (variable air volume) box 可变风量箱

VAV modular assembly 可变风量模块组件

VBR (variable bit rate) 变码率,可变比特率

VC (virtual circuit) 虚电路

VCI (virtual channel identifier) 虚拟信道标识符

VCS (video conference system) 视频会议系统

VCSEL (vertical cavity surface emitting laser) 垂直共振腔表面放射激光器

VDE (Verband Deutscher Elektrotechniker) 德国电气工程师协会

VDF (voice distribution frame) 语

音配线架

VDM (virtual DOS machine) 虚拟DOS机器

VDP (video door phone) 可视对讲系统

VDSL (very high data rate digital subscriber line) 甚高速数字用户线路

vehicle detection coil 车辆检测线圈

vehicle detection equipment 车辆检测设备

vehicle identification 车辆识别

vehicle information system (VIS) 车载信息系统

vehicle sensor 车辆感应器

vein recognition apparatus 静脉识别仪

velocity of light 光速

vented (secondary) cell 排气(二次)蓄电池

ventilation and air conditioning room 通风和空调机房

ventilation and air conditioning system 通风和空气调节系统

ventilation and heat abstraction 通风散热

ventilation hole 通风孔

Verband Deutscher Elektrotechniker (VDE) 德国电气工程师协会

verification 验证

vertical ancillary data (VANC) 垂直辅助数据

vertical built-in pipe 竖向暗配管

vertical cable trough 垂直理线槽

vertical cavity surface emitting laser (VCSEL) 垂直共振腔表面放射激光器

vertical distance 垂直距离

vertical hold 垂直同步

vertical interval switching 场间隔切换

vertical sampling 垂直采样

vertical scanning frequency (VSF) 场频

vertical shaft 垂直竖井

vertical system 垂直系统

vertical tilt 场倾斜

vertical viewing angle 垂直视角

very high data rate digital subscriber line (VDSL) 甚高速数字用户线路

very small aperture terminal (VSAT) 甚小孔径终端(系统)

vestigial sideband (VSB) 残留边带

vestigial sideband emission 残留边带发射

vestigial sideband filter (VSF) 残留边带滤波器

VFD (variable-frequency drive) 变

频器

VFR (voltage fluctuation rate) 电压波动率

VGA (video graphics array) 视频图形阵列

VGA distributor VGA 分配器

VGA interface VGA 接口

VHE (virtual home environment) 虚拟归属环境

VHS (video home system) 家用视频系统

VI (video interphone) 可视对讲机

vibration 振动

vibration frequency 振动频率

vibration isolation 振动隔离

vibration of cable 振动光缆

vibration sensor 振动传感器

video analysis module 视频分析模块

video analysis technology (VAT) 视频分析技术

video and audio transmission 视音频传输

video bandwidth 视频带宽

video cable 视频电缆

video camera 摄像机

video card 视频卡

video check to alarm 报警图像复核

video codec 视频编译码器

video compression standard 视频压缩标准

video conference system (VCS) 视频会议系统

video conferencing 视频会议

video conferencing technology 视频会议技术

video controller 视频控制主机

video detection 视频探测

video disk recorder 硬盘录像机

video distributor 视频分配器

video door phone (VDP) 可视对讲系统

video fast forward playback 视频快速前向重放

video frame rate 视频帧率

video gain adjustment 视频增益调节

video grabber 视频采集卡

video graphics array (VGA) 视频图形阵列

video home system (VHS) 家用视频系统

video image 视频图像

video indentification 视频识别

video input 视频输入

video input channel 视频输入通道

video input interface 视频输入接口

video intercom extension 可视对讲

分机

video intercom function 可视对讲功能

video interference 视频干扰

video interphone (VI) 可视对讲机

video interpolation 视频内插

video line 视频线

video loss alarm (VLA) 视频信号丢失报警

video matrix switcher (VMS) 视频矩阵切换器

video mixer 视频混合器

video moving detection 视频移动侦测

video on demand (VOD) 视频点播,交互式电视点播系统

video optical terminal 视频光端机

video output 视频输出

video payload 视频有效载荷

video printer 视频打印机

video projection 视频投影(机)

video signal 视频信号

video surveillance 视频监视

video surveillance and control system (VSCS) 视频安防监控系统

video switcher 视频切换器

video wall 视频墙

video wiper 视频消噪器

video format 视频格式

viewing angle 视角

viewing distance 视距

vignetting 渐晕

violation of traffic signal for power driven vehicles 违章机动车辆信号

VIPA (virtual IP address) 虚拟 IP 地址

virtual channel 虚拟信道

virtual channel identifier (VCI) 虚拟信道标识符

virtual circuit (VC) 虚拟电路

virtual connection 虚拟连接

virtual container 虚拟容器

virtual DOS machine (VDM) 虚拟 DOS 机器

virtual home environment (VHE) 虚拟归属环境

virtual instrument 虚拟仪器

virtual IP address (VIPA) 虚拟 IP 地址

virtual link 虚拟连接

virtual local area network (VLAN) 虚拟局域网

virtual memory (VM) 虚拟内存

virtual output queue (VoQ) 虚拟输出队列

virtual path (VP) 虚拟路径[通道]

virtual path identifier (VPI) 虚拟通路标识符

V

virtual private network (VPN) 虚拟专用网络
virtual reality (VR) 虚拟现实
virtual reality platform (VRP) 虚拟现实平台
virtual router redundancy protocol (VRRP) 虚拟路由冗余协议
VIS (vehicle information system) 车载信息系统
visible light communication (VLC) 可见光通信
visited location register (VLR) 访问地址寄存器
visitors intercom system (visual) 访客(可视)对讲系统
visual doorbell digital technology 可视门铃数字化技术
visual inspection 外观检查,目检
visual intercom 可视对讲
visual intercom function 可视对讲功能
visual verification 目视检查
visualization 可视化
VLA (video loss alarm) 视频信号丢失报警
VLAN (virtual local area network) 虚拟局域网
VLC (variable length code) 可变长度码
VLC (visible light communication) 可见光通信
VLR (visited location register) 访问地址寄存器
VM (virtual memory) 虚拟内存
VMS (video matrix switcher) 视频矩阵切换器
VOD (video on demand) 视频点播,交互式电视点播系统
voice communication 语音通信
voice distribution frame (VDF) 语音配线架
voice encoding 语音编码
voice encoding technique 语音编码技术
voice file 语声文件
voice over Internet protocol (VoIP) 网络电话,IP 电话
voice playback 语音播放
voice point 语音点
voice quickly connected plug 语音快接插头
voice synthesis (VS) 语音合成
voice-frequency circuit 音频电路
VoIP (voice over Internet protocol) 网络电话,IP 电话
voltage comparator 电压比较器
voltage fluctuation rate (VFR) 电压波动率
voltage grade 电压等级
voltage proof 耐压

voltage rating 额定电压
voltage stability 电压稳定性
voltage standing wave ratio (VSWR) 电压驻波比
voltage time integral variation 电压时间积分变化
voltage stabilizer 稳压器
volume 响度
volume capture ratio of annual rainfall 年降雨径流总量控制率
volume resistance 体积电阻
volume resistivity 体积电阻率
VoQ (virtual output queue) 虚拟输出队列
VP (virtual path) 虚拟路径，虚拟通道
VPI (virtual path identifier) 虚拟通路标识符
VPN (virtual private network) 虚拟专用网络
VR (variable resistance) 可变电阻
VR (virtual reality) 虚拟现实
VRP (virtual reality platform) 虚拟现实平台
VRRP (virtual router redundancy protocol) 虚拟路由冗余协议
VS (voice synthesis) 语音合成
VSAT (very small aperture terminal) 甚小孔径终端(系统)
VSB (vestigial sideband) 残留边带
VSCS (video surveillance and control system) 视频安防监控系统
VSF (vertical scanning frequency) 场频
VSF (vestigial sideband filter) 残留边带滤波器
VSWR (voltage standing wave ratio) 电压驻波比

W

WA (work area) 工作区

WAC (wide area centrex) 广域中心网,广域集中用户交换机业务

WAC (work area cord) 工作区跳线

WAIS (wide area information service) 广域信息服务

waiting space 等待[避难]空间

walking distance 步行距离

wall 墙壁

wall faceplate 墙面面板

wall-mounted distribution box 壁挂式配线箱

wall-mounted installation 壁挂式安装

wall mounting 墙面安装(支架)

WAN (wide area network) 广域网

WAP (wireless access point) 无线接入点

WAP (wireless application protocol) 无线应用通信协议

warded lock 弧形锁,凸块锁

warning device 威慑器

warning mark 警告标志

warranty of technical specification and performance of product 产品技术规格性能保证

waste 废弃物

waste component 废弃组件

waste material 废弃材料

water control valve (WCV) 水源控制阀

water curtain movie 水幕电影

water diff pressure switch (WDPS) 水差压开关

water flow indicator 水流指示器

water heater 热水器

water heating 水暖

water main 出水干管,总水管

water mist 水雾

water motor alarm 水力警铃

water penetration 渗水

water supply and drainage system 给排水系统

water to water plate heat exchanger 板式换热器

water transport factor of chilled water (WTFchw) 冷冻水输送系数
water-blocking tape 阻水带
water-blocking yarn 阻水纱
waterproof 防水
waterproof camera 防水摄像机
waterproof pigtail cable 防水尾缆
watt-hour meter demagnetization device 电能表退磁装置
wave splitter 分波器
waveform monitor 波形监视器
wavelength 波长
wavelength conversion module 波长转换模块
WBT (wet bulb temperature) 湿球温度
WC (wire cable) 线缆
WCDMA (wideband code division multiple access) 宽带码分多址
WCS (wireless communication system) 无线通信系统,无线通信网
WCV (water control valve) 水源控制阀
WD1 WD1 图像格式
WDLL (wireless digital local loop) 无线数字本地环
WDPS (water diff pressure switch) 水差压开关
WDR (wide dynamic range) 宽动态
WDR camera 宽动态摄像机
weak current system 弱电系统
weak system 弱电系统
weak well 弱电井
WEB report system (WRS) WEB 报表系统
WEB-based EPG 基于 WEB 的电子节目指南
weighting 计权
weld 焊接
welded electromagnetic shielding enclosure 焊接式电磁屏蔽室
welded steel pipe (WSP) 焊接钢管
welding auxiliary material 焊接辅助材料
wet 潮湿
wet alarm valve 湿式报警阀
wet bulb temperature (WBT) 湿球温度
wet contact 湿接(触)点
wet pipe system 湿式系统
WHA (whole home audio) 家庭背景音响
white balance 白平衡
white level 白电平
white light LED luminaire 白光 LED 灯具
white limiter 白色限制器
whole home audio (WHA) 家庭背

景音响

wide area centrex (WAC) 广域中心网，广域集中用户交换机业务

wide area information service (WAIS) 广域信息服务

wide area network (WAN) 广域网

wide dynamic range (WDR) 宽动态

wide screen projector 宽屏幕投影机

wideband code division multiple access (WCDMA) 宽带码分多址

wideband subscriber 宽带用户

wideband subscriber loop system 宽带用户环路系统

width height ratio 宽高比

width unit 宽度单位

wind velocity 风速

window signal 窗口信号

windows media audio (WMA) 语音编解码 WMA 音乐压缩格式

wire 电线

wire binding board 扎线板

wire binding hole 扎线孔

wire cable (WC) 线缆

wire casing 线槽

wire communication 有线通信

wire connector 导线连接器

wire duct 电线管

wire jumper 跨接线

wire laying 布线敷设

wire map 接线图

wire pair 线对

wire stripper 剥线钳

wire system 有线系统

wired broadcasting 有线广播

wireless access 无线接入

wireless access point (WAP) 无线接入点

wireless access unit 无线接入单元

wireless antenna 无线天线

wireless application 无线应用

wireless application protocol (WAP) 无线应用通信协议

wireless base station 无线基站

wireless channel 无线信道

wireless communication 无线通信

wireless communication system (WCS) 无线通信系统，无线通信网

wireless data communication equipment 无线数据通信设备

wireless digital local loop (WDLL) 无线数字本地环网

wireless door magnetic 无线门磁

wireless equipment 无线设备

wireless LAN (WLAN) 无线局域网

wireless local area network (WLAN)

无线局域网
wireless local loop (WLL) 无线本地环路
wireless mesh network (WMN) 无线网状网络
wireless microphone 无线传声器
wireless mode 无线方式
wireless network 无线网络
wireless paging system 无线呼叫系统
wireless projection 无线投影
wireless remote control 无线遥控器
wireless router 无线路由器
wireless sensor network (WSN) 无线传感器网络
wireless special-purpose chip 无线专用芯片
wireless telecommunications equipment 无线电信设备
wireless thermostat controller 无线恒温控制器
wireless transmission 无线传输
wireless walky-talky system 无线对讲系统
wire map 线路图
wiring 布线,接线,配线
wiring cabinet information 配线[布线]机柜信息
wiring closet 配线[布线]室
wiring design 配线[布线]设计
wiring diagram 配线[接线]图
wiring harness 导线束,把线
wiring inlet 导线入口
wiring management software 配线[布线]管理软件
wiring management system 配线[布线]管理系统
wiring optical cable 配线[布线]光缆
wiring order sheet 配线[布线]表
wiring patch information 配线[布线]架信息
wiring pipeline network 配线[布线]管网
wiring schedule 配线[布线]表
wiring subsystem 配线[布线]子系统
wiring system 配线[布线]系统
wiring terminal 配线[接线]端子
wiring zone 配线[布线]区
with shoulder bolt 阶式螺栓
WLAN (wireless LAN) 无线局域网
WLAN (wireless local area network) 无线局域网
WLL (wireless local loop) 无线本地环路
WMA (windows media audio) 语音编解码 WMA 音乐压缩格式

WMN (wireless mesh network) 无线网状网络
work area (WA) 工作区
work area cable 工作区电缆
work area cabling 工作区布线[布缆]
work area cord (WAC) 工作区跳线
work area subsystem 工作区子系统
work area wiring 工作区布线电缆
work order 工单
work station (WS) 工作站
working facility 工作设施
working principle 工作原理
working state 工作状态
workstation 工作站
worm gear reducer 蜗轮减速器
wrapping 绕包
wrinkle 皱纹
wrist strap 腕带,防静电手环
WRS (WEB report system) WEB报表系统
WS (work station) 工作站
WSN (wireless sensor network) 无线传感器网络
WSP (welded steel pipe) 焊接钢管
WTFchw (water transport factor of chilled water) 冷冻水输送系数
wye connection 星型[Y型]连接

X

XHTML (extensible hypertext markup language) 可扩展超文本标记语言

XML (extensible markup language) 可扩展标记语言

XMT (transmit) 发送,发射,传输,传播

X-ray X射线

XTLO (crystal oscillator) 晶体振荡器[谐振器]

Y

YCrCb　YUV 颜色空间

yellow-green wire　黄绿线

yoke vent pipe　结合通气管

YPBPR　模拟分量视频信号

YUV　YUV 信号

Z

Z 复阻抗

Z0 特性阻抗

ZBA (ZigBee building automation) ZigBee楼宇自动化

ZD (zero dispersion) 零色散

ZD (zone distributor) 区域配线架

ZDA (zone distributor area) 区域配线区

ZDS (zero dispersion slope) 零色散斜率

ZDW (zero dispersion wavelength) 零色散波长

zero dispersion (ZD) 零色散

zero dispersion slope (ZDS) 零色散斜率

zero dispersion wavelength (ZDW) 零色散波长

zero halogen 无卤素

zero line 零线

zero water peak optical fiber 零水峰光缆

ZigBee 紫蜂协议

ZigBee building automation (ZBA) ZigBee楼宇自动化

ZigBee channel ZigBee信道

ZigBee IP ZigBee IP协议

ZigBee PRO ZigBee PRO标准

ZigBee remote control (ZRC) ZigBee遥控

ZigBee RF4CE ZigBee RF4CE协议

ZigBee smart energy (ZSE) ZigBee智能能源

ZigBee2004 ZigBee 2004标准

ZigBee2006 ZigBee 2006标准

zinc die cast 锌压铸

zone 防区

zone distributor (ZD) 区域配线架

zone distributor area (ZDA) 区域配线区

zone fire alarm control unit 区域火灾报警控制器

zone management 分区管理

zoning override 分区强插

zoning with matrix mode 矩阵分区

zoom ratio 变焦倍率

ZRC (ZigBee remote control) ZigBee遥控

ZSE (ZigBee smart energy) ZigBee智能能源

Z-Stack Z-Stack协议栈

Z-Wave Z-Wave(由丹麦 Zensys 公司主导的无线组网技术)

以数字、希腊字母起首的词条

1080P (1920 × 1080 progressive scanning) 1080P(1920 × 1080 逐行扫描)图像格式

110 model distribution frame 110 型配线架

1280×720 progressive scanning (720P) 720P(1280×720 逐行扫描)图像格式

16 E1 interface board (GE16) GSM16 路 E1 中继接口板

1920×1080 progressive scanning (1080P) 1080P(1920×1080 逐行扫描)图像格式

2D (2-dimension) 二维

2-dimension (2D) 二维

2K 2K 图像格式(2048×1080)

32-bit microprocessor 32 位微处理器

3D (3-dimension) 三维

3D digital noise reduce (3D DNR) 3D 降噪

3D DNR (3D digital noise reduce) 3D 降噪

3-dimension (3D) 三维

4 pair balanced cable 4 对平衡线缆

4CIF 4CIF 图像格式

4K 4K 图像格式(4096×2160)

5G (5th-Generation) 第五代移动通信技术

64-bit extended unique identifier (EUI-64) 64 位扩展的唯一标识符

720P (1280 × 720 Progressive Scanning) 720P(1280 × 720 逐行扫描)图像格式

802.1x 802.1x 身份认证协议

960H 960H 图像格式(960×576)

λ0 零色散波长

λcc 光纤截止波长